Michael Bodendorfer

SWISSCASE

Michael Bodendorfer

SWISSCASE

Development and characterization of an electron cyclotron resonance ion source for the calibration of space flight instruments

Südwestdeutscher Verlag für Hochschulschriften

Imprint
Any brand names and product names mentioned in this book are subject to trademark, brand or patent protection and are trademarks or registered trademarks of their respective holders. The use of brand names, product names, common names, trade names, product descriptions etc. even without a particular marking in this work is in no way to be construed to mean that such names may be regarded as unrestricted in respect of trademark and brand protection legislation and could thus be used by anyone.

Publisher:
Südwestdeutscher Verlag für Hochschulschriften
is a trademark of
Dodo Books Indian Ocean Ltd., member of the OmniScriptum S.R.L Publishing group
str. A.Russo 15, of. 61, Chisinau-2068, Republic of Moldova Europe
Printed at: see last page
ISBN: 978-3-8381-0018-0

Zugl. / Approved by: Lausanne, Eidgenössische Technische Hochschule EPFL, Diss., 2008

Copyright © Michael Bodendorfer
Copyright © 2009 Dodo Books Indian Ocean Ltd., member of the OmniScriptum S.R.L Publishing group

Abstract

This PhD thesis consists of planning, simulating, building, testing and characterizing a new electron cyclotron resonance (ECR) ion source, **SWISSCASE**, the **S**olar **W**ind **I**on **S**ource **S**imulator for the **CA**libration of **S**pace **E**xperiments located at the University of Bern, Switzerland. The ion source will be operated in the existing CASYMS ultra high vacuum (UHV) facility and will extend the application field of the existing filament electron collision ion source inside CASYMS and the existing electron cyclotron resonance ion source MEFISTO, operated in its own UHV facility.

The chosen ECR ionization concept operating at 10.88 GHz delivers high currents of highly charged ions of up to 2 μA for Ar^{8+} and a maximum identified charge state of Ar^{12+}. In addition to argon, the ECR plasma of SWISSCASE has been operated with krypton, xenon and carbon dioxide gas, revealing all of the expected charge states and featuring a charge states distribution in favor of highly charged ions. Design constraints were given by the final application of SWISSCASE being implemented in CASYMS with limited space and power. The new ion source was realized with limited funding.

Numerical simulations gave insight in unprecedented quality and detail about the complex three dimensional extent of the magnetic field distribution of the full permanent magnet confinement of both, SWISSCASE and MEFISTO, the second ECR ion source, operated since 1997 at the University of Bern. For both confinement systems, the isocontour surface, defined by the electron cyclotron resonance condition of the incident microwave, is visualized in 3D.

In addition to the realization of the ECR ion source, a plasma Bremsstrahlung measurement revealed an average ECR electron temperature of 10 keV for an argon plasma. This temperature was used to perform numerical simulations of electron trajectories inside the SWISSCASE and the MEFISTO confinement. The simulated magnetic field distribution of MEFISTO and the simulated electron distribution of SWISSCASE are closely related to the observed triangle shaped surface coating patterns found in hexapole confined ECR ion sources.

Contents

1 Introduction   1
  1.1 Space exploration 1
  1.2 Space borne instruments 2
  1.3 Ground based ion sources to simulate the space environment . . 2

2 Requirements and selecton of ion source   7
  2.1 Introduction 7
  2.2 Situation and the need for a new ion source 7
  2.3 Requirements 9
  2.4 Assessed ion source concepts 10
    2.4.1 EBIS 10
    2.4.2 PIGIS 10
    2.4.3 ECRIS 11
  2.5 Choice of ion source concept 11

3 Electron cyclotron resonance   13
  3.1 Introduction 13
  3.2 Essentials 15
    3.2.1 Confinement single particle motion 16
    3.2.2 Microwave 19
    3.2.3 Vacuum 23
    3.2.4 Summary essentials 25
  3.3 Minimum-B ECRIS particle confinement 25
    3.3.1 Hot electron confinement 25
    3.3.2 Collision mechanisms and mean free paths 26
    3.3.3 Diffusion loss and ion confinement time 27
  3.4 Plasma potential 28
  3.5 Plasma density 30
  3.6 Comparison of SWISSCASE and MEFISTO 31
  3.7 Conclusion 31

4 Realization of source elements   35
  4.1 The magnetic confinement 36
    4.1.1 Overview 36
    4.1.2 Manufacturing process 37
    4.1.3 Assembly process 38
    4.1.4 Optimal setup 38
    4.1.5 Temperature regime of operation 40

	4.2	The microwave system	42
		4.2.1 The microwave generator	42
		4.2.2 Choice and characterization of the microwave generator	44
		4.2.3 Wave transport system and circulator	45
		4.2.4 Impedance matching	47
	4.3	Ion optics	47
		4.3.1 Extraction	48
		4.3.2 Baffle	50
		4.3.3 Einzel lenses	51
		4.3.4 Ion beam transmission	52
	4.4	Radiation shielding	53
	4.5	Vacuum setup and gas feed	55
		4.5.1 Vacuum capabilities	56
		4.5.2 Plasma chamber gas pressure	57
		4.5.3 Discussion	59
	4.6	The high voltage setup	60
		4.6.1 High voltage for extraction	61
		4.6.2 High voltage setup of ion optics	62
	4.7	Mass separation magnet	64
		4.7.1 Non relativistic justification	64
		4.7.2 Calculation of magnetic path integral	65
		4.7.3 Choice and purchase of mass separation magnet ...	66

5 Numerical Simulation of MEFISTO   67
5.1	Introduction	67
5.2	Finite element solver	68
5.3	Magnetic finite element model of MEFISTO	68
5.4	The ECR zone	71
5.5	Hot electron trajectories in MEFISTO	76
	5.5.1 Trajectory life times and lengths	77
	5.5.2 Discussion	81
	5.5.3 Relevance to laboratory plasmas	81
	5.5.4 Relativistic considerations	83
5.6	Conclusions of the MEFISTO simulation	84

6 FEM SWISSCASE   85
6.1	Introduction	85
6.2	Plasma Parameters	86
6.3	The magnetic field	88
6.4	The electron distribution	93
6.5	The observed surface coating pattern	95
6.6	Relation between simulation and observation	96
6.7	Conclusion	97

7 Source characterization and performance   99
7.1	Microwave coupling	99
	7.1.1 Frequency dependence	99
	7.1.2 Conclusion of the frequency dependence	103
7.2	Ion beam performance	104
	7.2.1 Measurement setup	104

		7.2.2	Faraday cup 105

       7.2.2   Faraday cup 105
       7.2.3   Argon spectra 109
       7.2.4   Krypton spectra 112
       7.2.5   Xenon spectra 114
       7.2.6   Carbon dioxide spectra 117
  7.3   Required ionization power 120
       7.3.1   Introduction 120
       7.3.2   Beam ionization power 120
       7.3.3   Discussion 121
  7.4   Summary of characterization 122

8 Bremsstrahlung measurement   **125**
  8.1   Introduction Bremsstrahlung measurement 125
       8.1.1   Need for high energy electrons 125
       8.1.2   Runaway electrons and non Maxwellian energy distribution 127
  8.2   Measurement setup 128
       8.2.1   The X-ray detector 129
       8.2.2   Calibration of X-ray detector 129
  8.3   Post processing 130
       8.3.1   Correction of attenuation effects 131
       8.3.2   Validation of the Bremsstrahlung spectrum 132
       8.3.3   Discussion of attenuation effect 134
  8.4   Determination of electron temperature 136
  8.5   Conclusion 137

9 Summary   **139**
  9.1   ECR Theory 139
  9.2   Realization of source elements 140
  9.3   FEM MEFISTO and FEM SWISSCASE 140
  9.4   Performance, characterization and fulfilled requirements 140
  9.5   Bremsstrahlung measurement 141
  9.6   Outlook 141
  9.7   Conclusion 142

10 Acknowledgments   **153**

Chapter 1

Introduction

1.1 Space exploration

The dominance of neutral matter on Earth is not representative for the rest of the solar system and the universe because space, as we know it, is dominated by vacuum and plasma rather than neutral matter [7]. The interstellar medium in general and the interplanetary medium of our solar system especially are fully ionized due to the long mean free paths, lack of recombination and the high energetic radiation environment of the solar wind and the cosmic rays background. Table 1.1 gives a summary of the sun's photosphere and the solar wind composition as an example of space particles found in the interplanetary medium.

Element	Rel. Abundance Photosphere $8+\log(n_x/n_H)$	Rel. Abundance slow solar wind n_x/n_O	Rel. Abundance fast solar wind n_x/n_O	Maximum charge state
H	8	1900 ±400	824 ±80	1+
He	10.93 ±0.004	75 ±20	48.5 ±5	2+
O	8.83 ±0.06	1	1	6+
C	8.52 ±0.06	0.72 ±0.1	0.70 ±0.1	6+
Ne	8.08 ±0.06	0.098 ±0.026	0.082 ±0.013	8+
N	7.92 ±0.06	0.129 ±0.008	0.145 ±0.011	6+
Mg	7.58 ±0.05	0.16 ±0.03	0.083 ±0.02	8+
Si	7.55 ±0.05	0.15 ±0.045	0.108 ± 0.023	9+
Fe	7.5 ±0.05	0.12 ±0.03	0.063 ±0.007	10+
S	7.33 ±0.11	0.05 ±0.018	0.053 ±0.013	9+
Al	6.47 ±0.07	0.0175 ±0.0066	0.008 ±0.002	11+
Ar	6.40 ±0.06	0.004 ±0.001	-	10+
Ca	6.36 ±0.02	0.017 ±0.003	0.0053 ±0.0014	10+
Na	6.33 ±0.03	0.0079 ±0.003	0.0038 ±0.0012	9+
Cr	5.67 ±0.03	0.0021 ±0.0004	0.0011 ±0.0002	10+

Table 1.1: Relative abundances and maximum charge states of the 15 most common elements in the solar photosphere, the slow and the fast solar wind [16].

Elements with larger atomic number are represented partially ionized with charge states up to 10+. An ion source used to calibrate space borne instruments for ion investigation has to reproduce these ions.

1.2 Space borne instruments

In order to explore space, instruments are developed to investigate the particle composition of virtually any matter found in the solar system. Neutral gas mass spectrometers investigate the composition of planetary and cometary atmospheres. Ion mass spectrometers explore the interplanetary media composed of the solar wind, coronal mass ejections and other sources of charged particles.

Mass spectrometers operate on charged particles, no matter whether the original particle is ionized or not. If neutrals are the particles of interest they get charged on entry to the mass spectrometer. Inside the mass spectrometers the charged particles are separated by energy, mass and charge by various means such as sector magnetic fields, quadrupole fields or time of flight measurements.

As an example of space borne particle instruments, Figure 1.1 shows both space borne PLASTIC [21] (**PLA**sma and **S**upra **T**hermal **I**on **C**omposition) mass spectrometers aboard the two STEREO (**S**olar **TE**rrestrial **RE**lations **O**bservatory) spacecrafts (each in one spacecraft). Both spacecrafts are simultaneously monitoring the sun from different orbital positions, providing a stereoscopic view of the sun. At the University of Bern, both instruments were tested and calibrated prior to flight in the MEFISTO and the CASYMS facility.

1.3 Ground based ion sources to simulate the space environment

Testing and calibrating space borne instruments such as PLASTIC requires ground based ion sources to simulate the space environment, where the instruments will be operated eventually.

Since free ions are unstable in the vicinity of neutral matter, appropriate high and ultra high vacuum facilities in combination with magnetic and electrostatic confinements are used to separate ions and neutrals and impede recombination. In addition, free ions do not occur naturally on Earth with very few exceptions such as in the Earth ionosphere. Ions in laboratories have to be created, confined and transported to the target with as little neutral matter interaction as possible.

Concepts of producing ions are numerous. Beside hot plasma methods and thermal ionization there are more subtle methods, more convenient for small space laboratory use. This is due to the different masses of ions and electrons. If a hot ion plasma is desired, ions of significant kinetic energy have to be confined in order to stabilize the plasma. However if hot plasma ions are not required per se and the mere charged ions are of interest, the plasma confinement can be realized in a far more economic way. Electrons are easier to confine than ions due to their smaller mass per charge ratio which constrains them more to electric and magnetic fields than ions.

Inside a plasma, more than one thermal population with different temperatures can coexist [7, 11]. Especially electron fractions and ions do not necessarily

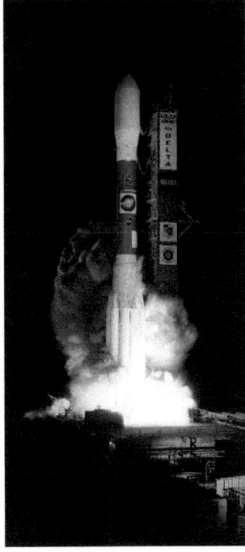

(a) PLASTIC instruments. (b) DELTA II launch.

Figure 1.1: (a) One PLASTIC instrument during the calibration procedure in the MEFISTO facility (upper image) and both PLASTIC instruments, after calibration, during the integration process (lower image). (b) Both PLASTIC instruments are part of the space mission STEREO, which was launched into space with a DELTA II rocket on October 26th, 2006

need to be part of the same temperature population. This plasma attribute is used to created high energetic electron population by various means and use them to ionize neutrals via electron impact ionization. In addition to this group of methods, there are other concepts of ionization, such as field ionization where locally very strong electric fields are used to disrupt neutrals to produce ions. However due to the large electric fields necessary for this methods the plasma sizes and total particle counts are relatively small compared to methods based on electron impact ionization.

Using electrons with a significantly larger kinetic energy than the background plasma ions brings the advantage of reduced recombination. This is due to the energy dependent recombination cross section of ions which diminishes for electrons in the range of several keVs [11, 29]. On the same hand, electrons in this energy regime can be confined about as well as ions in the range of a few eVs. Chapter 3 will further detail this phenomena. Figure 1.2 gives a comparison of size for a keV electron confinement (SWISSCASE) and a facility for keV to MeV ion confinement (JET). SWISSCASE (**S**olar **W**ind **I**on **S**ource for the **CA**libration of **S**pace **E**xperiments) is the electron cyclotron ion source realized within the frame of this thesis. JET (**J**oint **E**urpoean **T**orus) is the largest fusion research Tokamak in operation today.

(a) SWISSCASE confinement. Lower flange diameter: 250 mm. Permanent magnet mass: 18 Kg.

(b) JET facility (top) and torus view with plasma (bottom). Chamber height: 2.1 m (inner). Iron core mass: 2800 t.

Figure 1.2: (a): SWISSCASE confinement during assembly process. Permanent magnets are sufficient for the confinement of low density, anisotropically distributed ECR electrons of 10 keV. (b): JET for the magnetic confinement fusion research based at Culham, Oxfordshire, United Kingdom. Bigger and stronger solenoids are used for the confinement of higher density ions of 10 - 100 keV.

This suggests to set up an ion source with cold ions of small kinetic energy and hot electrons with large kinetic energy. Given the right parameters presented in this thesis, the plasma will be highly ionized and features ion energies far below their thermal ionization equivalent temperature, which can be found in the largest Tokamak magnetic fusion centers like JET (Joint European Torus, Culham, Oxfordshire, UK) and Tore Supra (Cadarache, Bouches-du-Rhone, France).

Following this way, the development and investigation of plasma based ion sources leads to electron cyclotron ion sources. Invented in 1965 ECR ion sources have continuously spread since then and can be found as work horses in ion implanters to the calibration of space borne instruments and up to high performance ion and neutral injectors for fusion Tokamaks with megawatts of injection power (JET, ITER).

All plasma based ion sources are limited in their ion performance by the number of available electrons with suitable kinetic energy. Plasma based ion sources with a single temperature plasma such as PIGIS and other arc based ion sources, the plasma electron density is always coupled to the neutral density. This is due to the plasma electron loss which needs to be compensated by fresh electrons originating from neutrals. However the requirement for a certain minimal neutral density advances recombination and thus limits the fraction of plasma ionization.

In contrast, ECR ion sources feature a highly anisotropic electron energy distribution due to the ECR heating process (further detailed in chapter 3). This pronounced anisotropy results in an excellent electron confinement exceeding by far the remaining plasma confinement performance. Hence a lower neutral density can still maintain a useful hot electron plasma density which would not be interesting in arc based ion sources. This combination leads to a very high ionization fraction and an unprecedented successful ion beam performance of ECR ion sources.

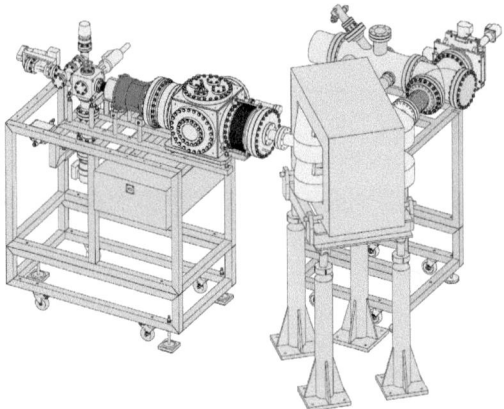

Figure 1.3: 3D Overview of SWISSCASE in green including the 90° mass separation magnet in blue and the Faraday cup for ion beam measurement and second vacuum pump facility in violet. The magnet confinement where the ECR plasma is located and the ion beam is formed is tinged orange.

To give an overview of the realized SWISSCASE ECR ion source Figures 1.3 and 1.4 show the whole facility in a 3D view and a top view respectively including the 90° mass separation magnet and the Faraday cup for ion beam measurement and a second vacuum pump facility.

In Chapter 2 the requirements of the desired ion source are presented and the concept selection is justified. Chapter 3 introduces the principle of operation of the chosen ECR ion source and details the involved single particle and plasma physics. Chapter 5 presents the magnetic model and the single particle simulation of the MEFISTO ECR ion source, also located at the University of Bern. Chapter 6 explains the results of the same kind of simulation applied to

6   Chapter: Introduction

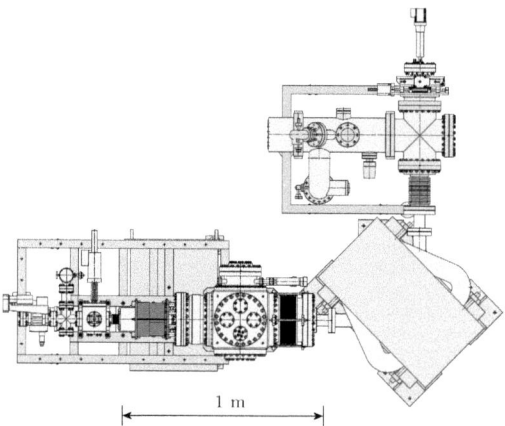

Figure 1.4: Top view of SWISSCASE in green including the 90° mass separation magnet in blue and the Faraday cup for ion beam measurement and second vacuum pump facility in violet. Color coding is identical to Figure 1.3.

SWISSCASE. Chapter 8 presents the results of a Bremsstrahlung measurement performed on SWISSCASE with the goal to access the plasma electron temperature. Chapter 9 summarizes the findings and points out suggestions for further investigations and improvements.

Chapter 2

Requirements and selecton of ion source

2.1 Introduction

In this chapter, the present situation is explained, which calls for a new ion source at the University Bern, for the calibration of space borne experiments. This leads to a list of requirements for the new ion source. Different ion source concepts are presented and evaluated. The concept of choice is justified and briefly characterized by its advantages and drawbacks.

For further concept details of the chosen electron cyclotron resonance ion source, please refer to Chapter 3. Chapter 4 gives details about the realization and Chapter 7 discusses the performance and characterization of the ion source.

2.2 Situation and the need for a new ion source

SWISSCASE has to fit into the existing CASYMS ultra high vacuum facility which is equipped with an electron collision ion source operating for more than 20 years already. SWISSCASE will complement the existing electron collision ion source rather than replace it. After the successful implementation of SWISSCASE there will be three fully operational ion sources at the University of Bern. SWISSCASE will be one of them. However the present filament based electron collision ion source and SWISSCASE will be operated in the same CASYMS facility whereas MEFISTO is operated in its own laboratory. The three ion sources will be used for different tasks, widening the total field of application:

- The **CASYMS** filament based electron collision ion source is used to simulate the Earth magnetosphere, cometary and planetary ionospheres and to the lesser extent, solar wind. It produces ion currents of singly and doubly charged ions of 10^{-8} A. The ion energy after an optional post acceleration ranges from 10 eV/q to 60 keV/q at present [21] and is planed to be increased to 150 keV/q after SWISSCASE is added to the facility. The ability for low energy ions down to 10 eV/q shall be retained after the SWISSCASE implementation.

- The **MEFISTO** 2.45 GHz ECR ion source is used to simulate the higher charge states of the solar wind, supra thermal ions and the interstellar medium. It produces ion currents of all charge states up to Ar^{8+} at low currents of 10^{-15} - 10^{-9} A. The ion energy can be varied from 3 keV - 100 keV. The lower limit is bound to the extraction voltage of the ECR ion source.

- The new ion source SWISSCASE is destined to simulate both lower and higher charge states of the solar wind, supra thermal ions, the interstellar medium and cometary atmospheres. It is supposed to produce all charge states up to Ar^{10+} with currents up to six orders of magnitude larger than MEFISTO and CASYMS. In addition it shall outperform the CASYMS filament electron collision ion source in terms of maintenance cycle time.

During a calibration process the space instrument is placed into the ion beam. The ion beam is wider than the entry aperture of the respective instrument. This leads to a geometric factor increasing the necessary ion beam current because a significant part of the ion beam is lost. In addition, many space instruments use channeltrons, an ion counter which is able to count single particles with a certain, limited frequency, typically 1 MHz. These constraints allow to calculate the ion beam current for every desired charge state, which would lead to channel-tron saturation and hence the shortest possible calibration time. Table 2.1 summarizes the resulting ion beam currents.

Charge state	Slit current	Total current
1+	$1.60*10^{-7}$ μA	$2.20*10^{-2}$ μA
2+	$3.20*10^{-7}$ μA	$4.45*10^{-2}$ μA
3+	$4.81*10^{-7}$ μA	$6.68*10^{-2}$ μA
4+	$6.41*10^{-7}$ μA	$8.90*10^{-2}$ μA
5+	$8.01*10^{-7}$ μA	$1.11*10^{-1}$ μA
6+	$9.61*10^{-7}$ μA	$1.34*10^{-1}$ μA
7+	$1.12*10^{-6}$ μA	$1.56*10^{-1}$ μA
8+	$1.28*10^{-6}$ μA	$1.78*10^{-1}$ μA
9+	$1.44*10^{-6}$ μA	$2.00*10^{-1}$ μA
10+	$1.60*10^{-6}$ μA	$2.23*10^{-1}$ μA

Table 2.1: Summary of currents received by an instrument slit of 12 mm x 15 μm (ROSINA), a circular beam diameter of 90 mm and a channeltron count rate of 1 MHz.

Table 2.1 shows that the total current is significantly larger than the slit current of a channeltron instrument if the maximal count rate of 1 MHz is used. Lower current and, hence lower count rates, result in longer exposure times. With the existing filament based electron collision ion source in CASYMS, the exposure times used to be as long as 10 hours. Such a long beam exposure time resulted in a change in beam current due to thermal drifts of the ion source. The table also shows total currents far larger than the currents extracted from MEFISTO. Because of the multidimensional parameter matrix of the space borne instrument, which has to be calibrated prior to instrument launch (energy, charge state, mass, incident angle, ect.) it also leads to calibration times which normally exceed the available project times by far.

New space instruments feature channeltrons and alternative particle counters with saturation count rates far beyond 1 MHz. This calls for an ion source with a total beam current orders of magnitude larger than the presently installed filament based electron collision ion source.

2.3 Requirements

The requirements for the new ion source are given by the situation as described in the previous section. Table 2.2 summarizes the requirements of the new ion source to give a quantitative overview.

Requirement	Quanta	hard/soft requirement
Ion charge states	Ar^{1+} to Ar^{10+}	hard
Ion charge states	Ar^{11+} to Ar^{n+}	soft
Ion current @ Ar^{8+}	0.178 μA	hard
Input power	< 750 W, single phase	soft
UHV capability at target	10^{-8} mbar	hard
Material cost	250'000 Swiss Francs	hard

Table 2.2: Summary of hard and soft requirements for the new ECR ion source in CASYMS.

The requirements are separated into hard and soft requirements. Hard requirements must be fulfilled at the exact specified value or better. Soft requirements can be fulfilled at the specified value, slightly worse or better. Partially fulfilled soft requirements do not disqualify the project but generally lead to higher cost in the following implementation and need to be avoided.

Table 2.2 points out that the available charge states of the new ion source have to reach at least up to Ar^{10+}. This is due to the maximum charge state present in the photosphere and the solar wind (see Chapter 1). Higher charges states than Ar^{10+} are desirable.

The ion current at the charge state Ar^{8+}, collected with a 10 mm aperture has to be at least $1.78 \cdot 10^{-1}$ μA. Ar^{8+} is the charge state of choice because it is the highest charge state, MEFISTO can produce for comparison. Because the new ion source will be operated on a high voltage potential (up to 150 kV), an isolation transformer is used to supply power to the ion source. The present insulation transformer has a spare capacity of 750 W. Exceeding this limit leads to the purchase of a new insulation transformer and the according costs.

The new ion source will be used to calibrate space flight instrumentation in ultra high vacuum of 10^{-8} mbar. Hence the new ion source must not contaminate the target chamber with ion source operation gas at a pressure higher than 10^{-8} mbar.

The budget for material cost of the ion source was limited to 250'000 Swiss Francs. It does not include labor costs or costs associated with material which is already in place at the University of Bern.

In the next section, the assessed concepts are presented to find the most suitable ion source type to be realized and implemented into CASYMS.

2.4 Assessed ion source concepts

Different ion source concepts have been assessed to guarantee an optimal choice. Available ion source concepts based on thermal ionization, spark discharge or laser ionization are not able to deliver the desired performance for the available funds. There are only three high performance ion source concepts of interest which are able to deliver highly charged ions at high ion currents which fulfill a majority of the remaining requirements too: EBIS, PIGIS and ECRIS. The three concepts have been evaluated in terms of usefulness, practicability, cost, development effort and ease of CASYMS integration.

2.4.1 EBIS

EBIS (**E**lectron **B**eam **I**on **S**ource) is an ion source concept based on a very dense and focused electron beam. The electron beam is generated by an electron gun and is accelerated by an electrostatic field to the desired energy. The electron beam collides with neutrals in a target vessel and produces ions. The advantage of this concept lies within the electron acceleration which is not limited by the operation parameters and is restricted only by the available space for the electron post acceleration. EBIS can produce the largest possible charge states of ions (U^{92+}, bare uranium, Lawrence Livermore Laboratory [25]). Despite its ability to produce highly charged ions, also this concept suffers from drawbacks:

- The electron beam can be used more efficiently by strong magnetic fields modifying the electron trajectories into helices and thereby increasing the effective path length inside the target gas. However this enhancement technique is limited and the electron beam finally leaves the target region, it is absorbed and lost. This leads to a rather low input power efficiency since most electrons of the electron beam fly either collision less or are scattered rather than causing ionization. The consequence is a large input power for the given performance requirement.

- Due to the electrostatic non-stochastic acceleration (as in ECRIS) the electrons feature a nearly mono-energetic velocity distribution. Due to the energy dependent ionization cross section of all neutrals, this may be suitable for the production of a single, even highly charged ion fraction but not if a wider variety of charge states is desired.

2.4.2 PIGIS

PIGISs (**P**enning **I**onized **G**auge **I**on **S**ource) use a hot electron emitting cathode in a hollow anode embedded in a static magnetic field. The emitted electrons are partially confined by the magnetic field and the opposite cathode allowing for many reflections before electron loss. Due to the reuse of plasma electrons, the Penning ion source is more efficient in terms of input power than the EBIS concept. However the PIGIS concept suffers inherently from drawbacks:

- Ion confinement times are low compared to ECRIS and cannot be improved by orders of magnitude necessary for competing with other concepts.

- The inherently high source pressure, necessary to sustain the arc discharge, supports recombination and impedes high charge states.
- The hot cathode suffers from well known life time limits ($\approx$ 20 h [11]). Electrode erosion during an operation cycle changes the ion source performance.
- The emitted electrons inside the arc are of low energy. Hence the ionization cross sections of highly charged ions are restricted.

2.4.3 ECRIS

ECRIS (**E**lectron **C**yclotron **R**esonance **I**on **S**ource) use an ECR plasma to ionize neutrals (see chapter 3). Early designs relied on solenoids to establish the magnetic confinement and the ECR process. Very significant progress in the past twenty years made permanent magnetic material available which features coercive field properties strong enough to deliver the desired magnetic field. First, the magnetic field for 2.45 GHz [16, 26], then 5.8 GHz, 7.5 GHz [36] and now for 10.88 GHz electron cyclotron can be supported by permanent magnetic material, without the need for solenoids and their associated power consumption. This allows the construction of an ECR ion source free of any power input except the microwave feed.

Because of the ECR process, electrons are very well confined due to their anisotropic energy distribution (see Chapter 3. Hence electrons which did not succeed in an ionization collision are recycled inside the ECR plasma for further electron impact ionization. The effective electron current of ECR ion sources used for ionization is many orders of magnitude larger than the electron current produced by filament based ion sources [11].

The ECR electrons establish a slightly negative plasma potential (-3 V) [14] and combine with the magnetic field to an ion confinement, superior to PIGIS. The ECR process and frequent collisions between ECR electrons result in a thermal electron distribution rather than a mono-energetic one [11, 3, 12, 19]. Overall, this leads to a very efficient ionization process with respect to power input, which is of highest interest for any ion source operated on a high voltage terminal.

2.5 Choice of ion source concept

Based of the summarized assessment, the ion source of choice is an electron cyclotron ion source (ECRIS). The chosen concept is characterized by advantages and their implications which other concepts do not benefit from:

- Most cost-efficient ion source at the specified performance requirements. While it may be possible to realize other ion source concepts with the same or better ion beam performance than an ECR ion source, an ECR ion source returns the best overall performance value for a given funding.
- Electrode-less operation is inherent for ECR ion sources since they do not require a hot filament or any other cathode. Consequently the ECR plasma can virtually be kept contamination free. In practice however wall sputtering and ablation are weakening this argument to a certain extent.

12   Chapter: Requirements and selecton of ion source

The lack of a corroding electrode eliminates electrode maintenance cycle time entirely. In addition, none of the three operation essentials, microwave, gas feed, magnetic field (see chapter 3 for further details) are degradable by the plasma [11].

The absence of an electrode furthermore allows the plasma operation with aggressive gases, such as O_2, F_2, Cl_2, which act highly consuming in filament based ion sources. The absence of a corrodible electrode makes the ECR ion source more reliable and stable with respect to ion beam performance.

- No need for any large power supplies (such as high power solenoid drivers) with full permanent confinements allows to integrate the ion source in a close space environment. This attribute is highly desirable for the room situation in the CASYMS laboratory.

- The ECR plasma can be operated at low pressures down to 10^{-5} mbar. The pressure immediately beyond the extraction assembly can be kept as low as 10^{-7} mbar and the final target chamber pressure (after the mass separation magnet) lies between 10^{-10} mbar 10^{-9} mbar [36]. This makes the ion source ideal for any ultra high vacuum (UHV) application.

- The thermal energy distribution of the ECR electrons produce a large variety of charge states from the same ECR plasma [11, 6, 36].

- The ECR concept promises low ion current noise if operated in quiescent and under dense plasma mode [11] (see Chapter 3).

Electron cyclotron resonance ion sources are the most attractive solution to satisfy the given demands because they combine a number of advantages with unprecedented reliability at the lowest cost rather than a single high performance record. However, as any other ion source concept also ECR ion sources present certain drawbacks which some other concepts do not inherently suffer from:

- Radiation. The ECR plasma produces X-ray radiation [36, 19]. Hence the ECR ion source needs X-ray shielding for operator safety.

- Temperature sensitive permanent magnet confinement further detailed in Chapter 7 changes the confinement and the ion beam performance.

- Minimum kinetic ion energy is bound to the extraction energy. ECR ions cannot be extracted at an arbitrarily low voltage without greatly impeding the overall ion source performance. Post deceleration of the ions is necessary but unattractive due to defocusing of the ion beam and subsequent ion beam current loss.

All these drawbacks can either be solved, circumvented, suppressed or improved to an acceptable level as will be shown in this thesis.

Chapter 3

Electron cyclotron resonance

In this chapter, the basic concept of electron cyclotron resonance (ECR) is explained. The essential elements of ECR are introduced and the resulting electron acceleration mechanism is discussed. A simple numerical simulation shows the universality of the ECR effect independent of the initial electron condition in velocity space. Collective plasma attributes are approached and limits of the current knowledge related to plasma diffusion and miminum-B confinements are presented.

Furthermore, the plasma density in SWISSCASE, a central plasma parameter, is detailed using data from Chapter 8, where the plasma temperature is derived from a Bremsstrahlung measurement. Finally, a comparison of the plasma parameters measured and calculated for SWISSCASE and MEFISTO reveals both, interesting similarities and differences, partially accounting for the gain in highly charged ion currents of SWISSCASE with respect to MEFISTO.

3.1 Introduction

Cyclotron resonance is a mechanism to accelerate charged particles in a magnetic field by way of resonantly coupling an oscillating electric or magnetic field to the charged particles. Beside few special application where oscillating and pumped magnetic fields are used, the majority of cyclotron resonance applications use oscillating electric fields. Despite the fact that oscillating electric fields are always accompanied by oscillating magnetic fields only the first is used for cyclotron resonance acceleration [11].

Particle accelerators, general purpose ion sources, ion thrusters and ion implanters further divide cyclotron resonance particle acceleration into ion cyclotron resonance (ICR) and electron cyclotron resonance (ECR). In ICR, the ions are directly accelerated by the resonance mechanism and can then be used for various applications such as further ionization, exhaust formation for space propulsion (ion-engine) or as ion beam injectors for nuclear fusion research. In ECR, electrons are accelerated in the first place, which then are used to ionize neutrals by electron impact ionization. Both using the same concept of resonance coupling, ICR generally operates at much lower frequencies than ECR due to the larger mass per charge ratio of ions compared to electrons. This follows from the resonance condition:

$$\omega_{ce} = \frac{qB}{m} \qquad (3.1)$$

ω_{ce} is also called the Larmor frequency. It describes the angular frequency at which any charged particle with mass m and charge q gyrates around magnetic field lines B. The resonance condition is generated by applying an external oscillating electric field with the same angular frequency ω_{mw} (see Figure 3.1). We have:

$$\omega_{mw} = \omega_{ce} \qquad (3.2)$$

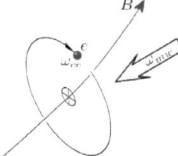

Figure 3.1: An electron moves on a circular path around a magnetic field line (blue) with an angular frequency ω_{ce}. Incident microwaves with the same angular frequency are directed toward the system.

This condition represents a harmonic oscillator in resonance. With no damping the energy of the particle in resonance would increase indefinitely. This is due to the fact that ω_{ce} is independent of the particle's kinetic energy. The resonance situation can be considered as undamped until either:

- the kinetic energy of the resonance particle gets close to its rest mass equivalent and special relativity has to be considered. This leads to a mass increase and hence a change of the resonance condition according to Eq.(3.1).

- or kinetic energies are large enough that cyclotron emission can no longer be neglected. This emission is due to the radial acceleration of the resonant particle.

In addition the resonance situation is limited by collisions with and scattering by other particles. Beside the emission of cyclotron radiation any of the other energy limiting mechanisms can be optimized to reach particle energies in the range of several MeVs. Next to ICR for industrial, scientific and space propulsion purposes, ECR is of highest interest for the production of ions. Despite the lower energy threshold where cyclotron emission and special relativity modifies the ECR condition, ECR is widely used to accelerate electrons up to energies large enough to ionize neutrals, singly charged ions and even completely strip the electron shells off an atom. In the medium energy range we find ECR ion sources which can produce any desired charge distribution.

The ECR process is randomized by collisions between ECR electrons and other plasma particles. This creates an energy distribution rather than a mono energetic electron population, resulting in a charge state variety rather than a single charge state domination. The variety of available charge states from the same continuous plasma enables ECR ion sources to satisfy not only industrial and commercial needs but especially represents an ion source for scientific applications of highest versatility.

The basic principle of ECR heating via microwaves enables an electrode-less operation, a tremendous benefit compared with electrode based ion sources such as EBIS. The absence of an electrode removes the operation cycle time limit restricted by electrodes of earlier designs. In addition the electrode-less operation of ECR ion sources allow an ECR plasma completely free of any material intrusion. This enables an ECR plasma free of contamination, an attribute highly desired in plasma physics of highly charged particles.

3.2 Essentials

To establish a successful ECR process three essential ingredients have to be in place.

- A magnetic field has to be present,
  - matching the resonance condition given by the microwave input frequency, electron mass and electron charge.
  - establishing the plasma confinement.
- An oscillating electric field has to be established to resonantly accelerate free electrons to kinetic energies large enough for ionization.
- A vacuum system has to guarantee mean free paths long enough:
  - for the acceleration process of the hot electrons.
  - to inhibit diffusion of both electrons and ions (see Section 3.3.3) and thereby enabling the magnetic confinement in the first place.
  - to inhibit recombination of highly charged ions.

Given these essentials an ECR ion source works by resonantly heated high energetic electrons colliding with slow neutrals or ions to increase their charge state. The electrons are confined in the ion source by the applied magnetic field. The magnetic field forms multiple magnetic mirrors (see section 3.2.1) resulting in a so called minimum-B field structure, a magnetic field configuration which features an increasing field in every direction moving away from the center of the ECR zone.

The ions are attracted by the magnetically trapped ECR electrons. Also ions undergo trajectory alteration due to the magnetic field. But due to their larger mass ions are less confined to the magnetic field than the more lightweight electrons of same energy. The charged particles inside a minimum-B configuration combined with a slightly negative plasma potential of minus 3 V, follow very complex particle trajectories not readily described or analyzed using analytical methods. However simplified models allow analytical insight and more sophisticated magnetic configurations can be studied by numerical finite element simulations.

The ionization charge state of neutrals or ions is increased by electron impact ionization, a process which removes an electron from an ion or a neutral by an impacting electron. The ionization probability depends on the particle charge state and the impacting electron energy. In addition to neutrals there is always more than one charge state present in an ECR plasma. In SWISSCASE an optimized argon ECR plasma consists of all argon charge states from Ar^{1+} to Ar^{12+}. Ionization cross sections are a function of the impacting electron energy. Ionization cross sections of higher charge states feature their maximums at higher electron energies [29, 24].

Unlike other ion sources such as electron beam ion sources, ECR ion sources feature an electron energy distribution rather than a mono energetic electron population. This leads to the mentioned variety of available charge states.

3.2.1 Confinement single particle motion

In this section, the first of the three ingredients, presented in the previous section, is introduced, the magnetic field for the particle confinement. To approach the concept of magnetic particle confinement, we start with single particle behavior in an inhomogeneous magnetic field forming a magnetic bottle, the most simple case of a magnetic confinement. Later in this section we related the magnetic properties of the confinement to the second of the three ingredients, the microwave heating, which is detailed in the next section.

To investigate the trajectory of an electron trapped by magnetic field we consider the following. In a collision-less plasma charged particles gyrate around magnetic field lines with the Larmor radius ($r_L = vm/qB$) given by the particles velocity v, mass m, charge q and the magnetic field strength B. If an additional longitudinal velocity component with respect to the local magnetic field is present, the whole motion is a helix along magnetic field lines.

To establish a magnetic bottle confinement as used in ECR ion sources a inhomogeneous static magnetic field is used. The particle trajectory, the helix, changes as soon as the particle enters a volume of different magnetic field. There are two invariants which allow to determine the continuing trajectory of the particle. Both the kinetic energy of the particle E_{kin} and the adiabatic invariant μ are conserved [7]:

$$E_{kin} = \frac{1}{2}mv^2 \qquad (3.3)$$

$$\mu = \frac{1}{2}\frac{mv_\perp^2}{B} \qquad (3.4)$$

m is the particle mass, v its velocity, $v_\perp$ its velocity component perpendicular to the magnetic field B. Since $v^2 = v_\perp^2 + v_\parallel^2$ we can combine 3.3 and 3.4:

$$\frac{2E_{kin}}{m} = v_\parallel^2 + \frac{2\mu B}{m} \qquad (3.5)$$

From Eq.(3.5) it is clear, any increase in B leads to a decrease in $v_\parallel^2$ to keep the left side of Eq.(3.5) constant. Since the kinetic energy (3.3) is invariant, the decrease in $v_\parallel^2$ must lead to an increase in $v_\perp^2$:

$$v_\perp^2 = \frac{2E_{kin}}{m} - v_\parallel^2 \tag{3.6}$$

In the most simple case a magnetic bottle consists of a volume with B_1 and B_2 and $B_2/B_1 = r_m > 1$ (see Figure 3.2).

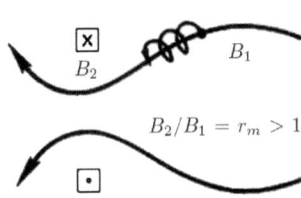

Figure 3.2: The most simple case of a magnetic mirror: the magnetic field is distorted by a solenoid (see square cross section with dot and cross for current indication), increasing the magnetic field at B_2 with respect to B_1. The particle moves on a helix along the magnetic field lines. The particle is confined, if its initial velocity at the location of B_1 fulfills the mirror criteria $v_{\parallel 1}^2/v_{\perp 1}^2 > B_2/B_1 - 1$ (see Eq. 3.9).

r_m is called the mirror ratio of the magnetic bottle if B_2 is the maximum field density and B_1 the minimum field density. The Larmor radius gets smaller as $v_\perp^2$ increases (3.6) and the Larmor frequency gets larger. Additionally, this effect can also be explained by a different approach, introducing the magnetic flux Φ_A enclosed by the electron's circumferential motion path:

$$\Phi_A = \int_{\partial A} \vec{B} \bullet \vec{dA} \tag{3.7}$$

We assume $\vec{B}$ to be constant in the plane of $\vec{A}$: $\Phi_A = \vec{A} \bullet \vec{B}$. We can further use the Larmor radius r_L to express $|\vec{A}| = r_L^2\,\pi$. Hence:

$$\Phi_A = r_L^2 \pi \frac{2\,\mu\,m}{q^2\,r_L^2} = \frac{2\pi\,\mu\,m}{q^2} = const. \tag{3.8}$$

This means that an electron circling along its helix toward or away from the magnetic mirror maximum keeps enclosing a constant magnetic flux by its Larmor radius defined circular motion (see Figure 3.3).

Further considerations lead to the confinement performance of the magnetic bottle. For this we assume a particle to start at the location with B_1, the field minimum, traveling along the magnetic field toward B_2. The particle may have an initial velocity $v_1^2 = v_{\perp 1}^2 + v_{\parallel 1}^2$. As the particle travels toward B_2, the shape of its motion helix changes (see Figure 3.3). When the particle finally arrives at B_2 it has a velocity $v_2^2 = v_{\perp 2}^2 + v_{\parallel 2}^2$. We further require the particle to be just confined ($v_{\parallel 2}^2 = 0$) rather than to over shoot the magnetic field maximum. Further simplification leads to:

$$\frac{v_{\parallel 1}^2}{v_{\perp 1}^2} + 1 = \frac{B_2}{B_1} \tag{3.9}$$

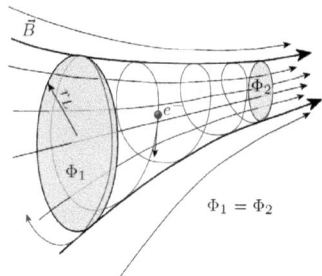

Figure 3.3: An electron moves on a helix along the magnetic field lines toward or away from the magnetic mirror maximum. The circumferential path describes a circle enclosing an area (light blue) which is changing with the distance from the field maximum. The enclosed magnetic flux Φ is invariant. The Larmor radius r_L decreases during the particle's motion toward the field maximum and increases on its way back.

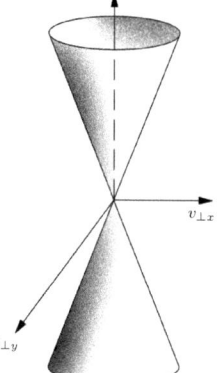

Figure 3.4: Loss cone. All particles with $v_{\|1}^2/v_{\perp 1}^2 > 1$ are lost from the magnetic confinement bottle. This criteria represents a cone shape in velocity space. All electrons with velocity vectors inside the cone are lost. Electrons with velocity vectors outside the cone are confined by the concept mirror.

(3.9) defines the so called loss cone (see Figure 3.4). The loss cone is a volume in velocity space given by the condition that all particles with $v_{\|1}^2/v_{\perp 1}^2 > B_2/B_1 - 1$ are lost from the confinement.

The ratio $v_{\|1}^2/v_{\perp 1}^2$ significantly differs for ions, cold electrons and hot electrons. Due to thermalization ions and cold electrons share the same temperature. Despite the presence of a magneto-static field they feature an approximately isotropic velocity distribution [11] and feature a similar $v_{\|1}^2/v_{\perp 1}^2$ ratio close to 1. This is very different for the hot electron population. Simplified the hot electrons are only heated perpendicularly to the magnetic field lines due to the electron cyclotron resonance mechanism presented in the following section. This leads to a strong anisotropy in the velocity distribution because the parallel velocity component remains low with energies comparable to the cold electrons and ions, whereas the perpendicular velocity component can reach energies energies from 1.7 keV (MEFISTO [17]) up to several hundred keV [11]. In SWISSCASE this velocity component has an average energy of 10 keV (see chapter 8 Bremsstrahlung).

For the hot electron population, $v_{\|1}^2/v_{\perp 1}^2$ is roughly 2eV/10keV = $2 \cdot 10^{-4}$.

This allows to simplify the plasma situation considerably by stating that the hot electrons are perfectly confined unless they undergo a collision or scattering. This brings the mean free path and the collision frequencies between hot electrons and neutrals and ions to our attention, further detailed in the following section.

For the successful operation of an ECR plasma a high magnetic field is highly desirable since it defines the microwave frequency (see Eq. 3.1 and 3.2). This in turn puts an upper limit on the plasma electron density (see next paragraph), desired to as high as possible and necessary for high currents of highly charged ions. According to Geller [11], the following reasoning can be stated. Note the square root relation between ω_{pe} and I_q, which makes a higher operation frequency ω_{ECR} very attractive:

$$B \sim \omega_{ECR} > \omega_{pe} \sim \sqrt{I_q} \qquad (3.10)$$

B is the magnetic field, ω_{ECR} the angular frequency of the microwave input feed, ω_{pe} the plasma electron angular frequency and I_q the extracted current of charge state q. This reasoning requires operating an under-dense plasma. This means the ECR plasma becomes opaque for the incident microwave when the plasma electron density reaches its critical value and the corresponding plasma frequency cuts off any incident microwaves [7] (see Section 3.22 for more detail). However, also over-dense ECR plasmas can be created by microwave coupling via X-mode. Over-dense ECR plasmas tend to be unstable and feature a high effervescence not desired for the extraction of a low noise ion beams and are therefore avoided.

In practice B_2 is limited by the chosen concept or material. B_1 however can be arbitrarily designed to any value between 0 and B_2. The lower B_1, the better the mirror ratio r_m. The higher B_1 the better the electron density and the ion charge state distribution due to a higher ECR frequency. For SWISSCASE a trade off is chosen with $r_m \approx 2$ (see Ph.D. thesis by Trassl [36], 1999).

The confinement of SWISSCASE consists of an axial magnetic mirror and a Halbach hexapole. The Halbach hexapole represents six equally spaced magnetic mirrors oriented radially rather than axially (see figure 4.4b, chapter *realization*). Magnetic fields can be superimposed to access the resulting magnetic field of both the axial mirror and the Halbach hexapole radial mirrors. This allows to realize a magnetic confinement with an increasing magnetic field in every direction from the center of the confinement, a minimum-B confinement. However for this kind of magnetic confinement, particle trajectories calculations are far from being solvable analytically as shown for the simple magnetic mirror. The results of 3D numerical trajectory integrations in the SWISSCASE minimum-B confinement are presented in chapter 6.

In the next section we go into more detail about the second necessary ECR ingredient, the microwave input feed.

3.2.2 Microwave

As shown in the previous section charged particles gyrate around magnetic field lines with a frequency given by $f_{ECR} = \frac{qB}{m}/2\pi$. This frequency does not depend on the energy of the particle.

When a charged particle is trapped by a magnetic field line it gains kinetic energy and its Larmor radius increases to match the higher velocity. This situation is highly useful since it guarantees the Larmor frequency to be the same for all trapped electrons in a constant magnetic field, independent of their energy.

An electric field as part of an electro magnetic wave hitting a particle circling a static magnetic field line can be decomposed into an electric field component oscillating parallel to the magnetic field line and another component oscillating perpendicular to the magnetic field line. The field component oscillating parallel to the magnetic field line interacts with the trapped electron but does not lead to an energy increase averaged over time due to the nearly linear oscillating motion along the field line. In contrast, the component perpendicular to the magnetic field line and hence parallel to the electron gyro motion does interact with the electron changing the electrons kinetic energy averaged over many gyro rotations if the incident wave matches the electron cyclotron frequency.

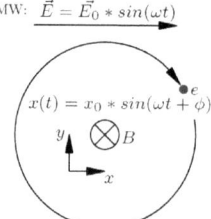

Figure 3.5: The incident microwave (top) supplies the gyrating electron with a homogeneous oscillating electric field. Before field excitation the electron features a circular trajectory around a magnetic field line resulting in a harmonic function for its x location including a phase shift ϕ with respect to the incident microwave.

To investigate this acceleration mechanism we introduce figure 3.5. The electron trajectory in the plane perpendicular to the magnetic field depends on the phase shift between the electrons initial oscillation and the driving electric field of the incident microwave. Figure 3.6 shows a numerical trajectory integration for four different phase shifts. The simulation is accurate to 1% of the electron velocity within 110 cyclotron orbits or 10^{-8} s, the time required to obtain 10 keV (see below). A homogeneous electric field is assumed. For SWISSCASE this assumption is justified by the small ratio between the larmor radius of the electrons (0.868 mm) and the wavelength of the incident microwave (27.55 mm). Simplifying the total microwave generator output power of 100 W to be adsorbed by the ECR zone cross section allows to calculate the resulting electric field E_0:

$$E_0 = \sqrt{\frac{P_{mw}}{\epsilon_0 \, c \, A}} \qquad (3.11)$$

P_{mw} is the microwave generator output power, c the speed of light in vacuum and A the cross section of the ECR zone. For SWISSCASE values we obtain $E_0 = 18$ V/mm. This value for the accelerating electric field $E(t) = E_0 sin(\omega t)$ is chosen for all simulation results presented in this section. The initial kinetic electron energy is set to 2 eV, the temperature of the cold electron population.

The simulation takes into account the presence of the oscillating electric field and a static, homogeneous magnetic field rather than a mirror field. It

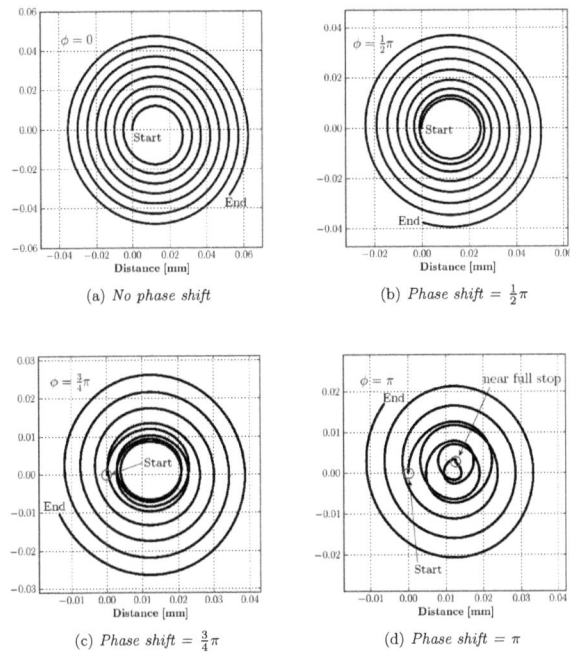

Figure 3.6: Electron trajectories in resonance with a magnetic field $B = 0.388$ T and an oscillating electric field with $E_0 = 18$ V/mm in x direction. Four different phase shifts ϕ between the electron initial oscillation and the driving electric field from 0 to π are shown.

does not take into account any effect related to scattering, collisions, cyclotron emission or other dissipative damping mechanism. The simulation is realized by a simple loop integrator with a time step of $6.6 \cdot 10^{-14}$ s, 10^5 integration steps and a total time span of 10^{-8} s. This corresponds to $7.25 \cdot 10^{-4} \cdot \tau$, with τ being the period of the cyclotron motion. In each integration step, the loop integrator first computes the Lorentz force due to the magnetic field, the force caused by the electric field of the incident microwave, the resulting acceleration, the new velocity and the new location based on the previous time step. This cycle is repeated until the preset loop number is reached and the simulation is terminated.

Figure 3.6 shows that the early part of the electron trajectory significantly depends on the phase shift between the initial electron motion and the incident microwave. A red circle in Figure 3.6d, showing the scenario with a phase shift

$\phi = \pi$, marks the trajectory point where the electron undergoes a near full stop to synchronize with the driving electric field.

All four scenarios show a similar resonant spiral after several rotations. The initial phase shift hardly matters for electron energies in the range of several keVs. To visualize this effect Figure 3.7 shows the development of the kinetic electron energy with respect to time for larger numbers of rotations.

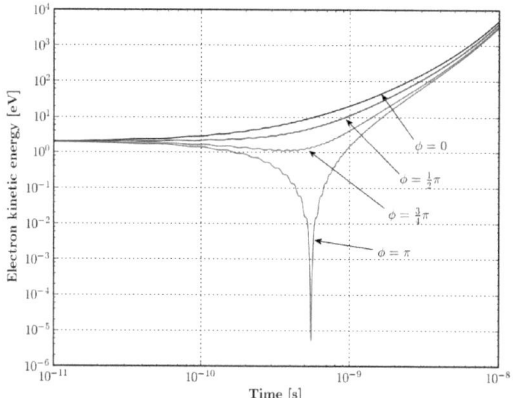

Figure 3.7: Kinetic electron energy with respect to time for different initial phase shifts. Ripples in the electron energy attest to the periodic energy input from the oscillating driving electric field. Note the low energy spike of the scenario for $\phi = \pi$ in green representing a near full stop. All four scenarios show a very close energy development for energies in the range of keVs.

Figure 3.7 shows that the early kinetic energy development of the resonant electron significantly depends on the initial phase shift of electron motion and incident microwave, as already discovered in Figure 3.6. The scenario with $\phi = \pi$ (in green) features a low energy spike, representing the near full stop necessary to synchronize the phase of the electron motion and the driving electric field. In comparison the electrons of the scenarios with $\phi = \frac{1}{2}\pi$ (in blue) and $\phi = \frac{3}{4}\pi$ (in magenta) get decelerated a lot less before the phase locked acceleration. However approaching kinetic energies of several keVs all scenarios close in on similar electron energies.

Figure 3.8 displays the kinetic energy an electron gains with respect to the distance it has traveled in the scenario. Again results are shown for the same four different phase shifts presented in figure 3.6 and 3.7. The electrons gain a kinetic energy of 10 keV within 108 to 110 cyclotron orbits and within a distance of 0.27 m to 0.372 m respectively. This distance is well within the mean free path of the electrons (4.9 m) presented further down this section.

The similar behavior for different initial phase shifts is a very important effect of electron cyclotron resonance because it synchronizes the electrons in

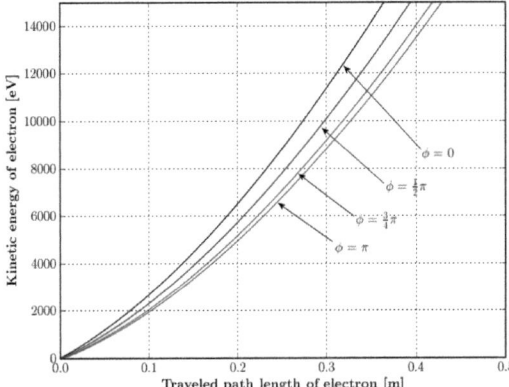

Figure 3.8: Kinetic electron energy with respect to the distance the respective electron has traveled. The scenarios for four different initial phase shifts are presented. The electrons reach an kinetic energy of 10 keV after 0.27 to 0.33 m depending on the initial phase shift.

energy space. Hence electrons with chaotic initial conditions are accelerated to similar energies and initial phase shifts do not matter for the energy distribution of the hot electron population.

Another robust attribute of ECR electron heating by microwaves is the geometric tolerance of the ECR effect. The sophisticated spatial orientation of the local magnetic field lines guarantees a non zero electric field component perpendicular to the magnetic field lines for almost all locations. Only electrons attached to magnetic field lines oriented exactly parallel to the oscillating electric field (O-wave microwave mode) do not benefit from the resonance effect.

3.2.3 Vacuum

Any electron acceleration process such as the electron cyclotron resonance mechanism, presented in the previous section, requires a certain mean free path for the electrons to complete the acceleration up to energies high enough for ionization. The presented numerical simulation has shown that this mean free path needs to be at least 0.33 m for the ECR process realized in SWISSCASE. In this section the necessary vacuum condition will be presented to enable such a situation.

In a collision-less plasma the fraction of particles inside the loss cone would be lost and the remaining particles would be confined indefinitely. However real plasmas feature and ECR plasmas even require collisions to produce the desired ions. Collisions in turn lead to recombination, scattering into the loss cone (loss cone scattering) and diffusion across magnetic field lines, an additional confinement loss poorly understood today. Hence a trade off has to be found to allow enough collisions for a most effective ion production but to inhibit unnecessary collisions to minimize recombination, loss cone scattering and diffusion loss of both electrons and ions.

The SWISSCASE plasma chamber pressure, when used for ion beam production, lies within a pressure range of $5 \cdot 10^{-5}$ mbar to $8 \cdot 10^{-4}$ mbar. The exact temperature of the neutral particles in the SWISSCASE plasma chamber is unknown. The lower limit for the temperature is given by the plasma wall temperature. However the neutrals undergo frequent collisions with ions and electrons of higher temperature [7, 11, 17] which leads to an increase in neutral temperature. In addition some neutrals are the result of a successful recombination process of singly charged ion and an electron, both having a higher temperature than the surrounding neutrals.

This reasoning suggests the neutrals feature a temperature in between room temperature (plasma chamber walls) and 2 eV [17], the temperature of the cold electrons. To calculate such important plasma parameters as mean free path lengths and density we need to assume a temperature of the neutral particles. According to the ideal gas law the lowest possible temperature leads to the highest possible particle density and the shortest mean free paths. Consequently we will calculate all parameters of the neutral population with a neutral temperature of $T_n = 300$ K to obtain minimal values for the mean free paths.

The mean free path λ_{free} in a particle ensemble of density n and particle collision cross section σ is given by [7]:

$$\lambda_{free} = \frac{1}{n\,\sigma} \qquad (3.12)$$

CO_2 gas contains oxygen and carbon which are of highest interest for the calibration of space borne instrumentation. CO_2 is a straight molecule with no polarity. We take a collision cross section of $4.25 \cdot 10^{-20}$ m^2 for an electron hitting a CO_2 molecule (see Chapter 6 for details). With $T_n = 300$ K we get a CO_2 neutral density of $4.8 \cdot 10^{18}$ 1/m^3. In an ECR plasma there are not just CO_2 neutrals but also other neutrals such as C and O radicals and CO and O_2 molecules. To summarize this density multiplying effect we simply double the density to compensate for split up CO_2 molecules (priv. com. Prof. Wurz, February 2008). Implementing these values into formula (3.12) results in λ_{free} = 4.9 m at a pressure of $2 \cdot 10^{-4}$ mbar which is the mean optimal pressure for CO_2 plasma operation in SWISSCASE. Hence the mean free path, limited by the neutral density, is long enough to allow the build up of electron energies high enough for ionization.

At a plasma chamber pressures of $2 \cdot 10^{-3}$ mbar or higher the SWISSCASE plasma ceases to output any high charge state ions. Applying the simplified ideal gas model for this pressure we obtain a mean free path of 0.49 m which is very close to the 0.33 m necessary for the cyclotron acceleration, shown in the previous section. Hence the numerical simulation result presented in the previous section is in good agreement with the simplified ideal gas model for the neutral density.

3.2.4 Summary essentials

The electron cyclotron resonance acceleration mechanism can be explained by single particle motion and numerical integration of the electron trajectories. The described acceleration process is robust and works for a thermal ensemble of electrons with isotropic velocity distribution.

3.3 Minimum-B ECRIS particle confinement

To obtain singly charged ions from an ECR ion source, the ions do not need to be confined by the ion source because they can be extracted right after the ionization process. This is very different for ECR ion sources aiming at multiple charged ions.

The ionization of a multiple charged ion can be performed by the removal of one electron per ionization step or by the removal of multiple electrons in the same ionization step. The effective cross section for the removal of two electrons is generally two orders of magnitudes lower [33] than the effective cross section for the removal of a single electron per ionization step. Consequently the production of multi-charged ions is dominated by a step by step process where a single electron is removed per step [17].

Hence each ion with charge state N needs to undergo an ionization at least N times without recombination. With recombination present within the sequence of ionization the total number of successful ionization steps is even higher.

Compared to single or low charge state ions, highly charged ions require:

- longer confinement times to enable multiple ionization events.

- shorter mean free path lengths with respect to hot electrons to increase hot electron collision frequency and decrease the necessary confinement time

- longer mean free path lengths with respect to neutrals to inhibit recombination.

- lower overall plasma density to inhibit malignant charge exchange.

This suggests an increasing microwave power input to increase the hot electron population and a decreasing optimal feed gas pressure to lower the neutral density for optimal production of highly charged ions. SWISSCASE confirms the expected behavior by a monotonously increasing high charge state ion output with monotonously increasing microwave input power and monotonously decreasing feed gas pressure down to $5 \cdot 10^{-5}$ mbar.

3.3.1 Hot electron confinement

Hot electrons are well outside the loss cone defined by the simple magnetic mirror and confined indefinitely due to the pronounced anisotropic velocity distribution in favor of the perpendicular velocity component with respect to the magnetic lines. Hence to loose hot electrons their anisotropic velocity distribution has to be altered in a way to expel the electrons, capture them or decrease their kinetic energy enough to disqualify them as hot electrons. The dominant mechanisms of this alteration are collisions and scattering.

In addition to loss cone scattering, collision absorption and recombination, hot electrons can diffuse across magnetic field lines by scattering. Diffusion across magnetic field lines is considered to finally lead to electron loss [11] which is why this type of diffusion is a very important confinement parameter.

3.3.2 Collision mechanisms and mean free paths

Any plasma consists of charged particles which may feature a significant collision cross section for electrons. This is due to the electric field originating from any charged particle. Due to their mass, ions are slow relative to both the hot electrons and the cold electrons and are considered immobile in the following discussion.

An electron passing by an ion is scattered by the ions electric field. According to Chen [7] and Geller [11], an oder-of-magnitude estimate for the change in momentum of a scattered electron can be obtained by assuming a 90° deflection.

$$\Delta p_e = m_e \, \Delta v_e = \frac{e^2}{4\pi \, \epsilon_0 \, r_0 \, v_e} \qquad (3.13)$$

Δp_e is the change in electron momentum, e its charge, v_e its velocity and r_0 the so called impact distance, the projected distance of the initial electron velocity vector toward the ion. In a 90° scattering event the initial momentum of the electron is completely lost with respect to the direction of initial motion. Hence $\Delta p_e = p_e = mv_e$. Given this we can solve for the impact distance which represents the radius of a circular scattering cross section σ_{ei}.

$$r_0 = \frac{e^2}{4\pi \, \epsilon_0 \, m_e \, v_e^2} \quad => \sigma_{ei} = \frac{e^4}{16\pi \, \epsilon_0^2 \, m_e^2 \, v_e^4} \qquad (3.14)$$

Note the occurrence of the fourth power of v_e. This leads to a very strong decrease of σ_{ei} with increasing v_e. The decreasing collision cross section with increasing electron velocity and energy is characteristic for all momentum exchange effects involving long range forces such as electric or magnetic fields.

In addition to the described one time 90° deflection, the sum of multiple small angle scatterings are significant, called Spitzer collisions. The total collision cross section from Spitzer collisions turns out to be [11]:

$$\sigma_{eispitzer} = \left(\frac{e^2}{\epsilon_0 \, E_{kin}}\right)^2 \frac{1}{2\pi} \, \ln \Lambda \qquad (3.15)$$

which is a factor of $32 \ln \Lambda$ larger than σ_{ei}. $\ln \Lambda = 24.62$ is the Coulomb logarithm for SWISSCASE, weakly depending on the plasma parameters. Hence the limiting cross section is dominated by multiple small angle collisions rather than large angle close encounters. For a SWISSCASE plasma we can summarize the values of these two collision mechanisms in table 3.1.

In comparison to the values for one time 90° and multiple small angle Spitzer collisions shown in table 3.1, the collision parameters of 10 keV hot electrons with neutrals are dominant. This is an important conclusion used in the following to calculate the diffusion loss rates across magnetic field lines.

Mechanism	σ	λ_{free}	ν
one time 90° [7]	$3.2 \cdot 10^{-26}$ m^2	$2.1 \cdot 10^7$ m	2.8 Hz
multiple small angle Spitzer [11]	$1.2 \cdot 10^{-23}$ m^2	$5.6 \cdot 10^4$ m	$1.0 \cdot 10^3$ Hz
Neutral collisions CO$_2$	$4.2 \cdot 10^{-20}$ m^2	4.9 m	$1.2 \cdot 10^7$ Hz

Table 3.1: Summary of cross sections σ, mean free paths λ_{free} and collision frequencies ν of hot electrons (10 keV) for one time 90° and multiple small angle Spitzer collisions. Hot electron density: $1.46 \cdot 10^{18}$ 1/m^3, neutral density: $4.8 \cdot 10^{18}$ 1/m^3 and plasma chamber pressure: $2 \cdot 10^{-4}$ mbar.

3.3.3 Diffusion loss and ion confinement time

Diffusion loss rates across magnetic field lines is a subject not yet accessible in an analytical way [7, 11]. Empirical and semi empirical formulas for the calculation of the diffusion coefficient D show that classical, analytical approaches are too optimistic in favor for the confinement performance by many orders of magnitude. On the other hand, Bohm diffusion suggests a diffusion coefficient many times too pessimistic to be of use for mininum-B ECRIS confinement engineering [11].

For hot electrons, diffusion is useful to approximate the number of collisions, necessary to let an electron collide with the plasma chamber wall. This calculation is detailed in chapter 6. However for the ECR performance, hot electron diffusion is of little interest because diffusion relies on successive collisions. As stated above we consider a hot electron as energetically degraded and uninteresting for ionization if it undergoes a single collision.

Cold electrons and ions however would potentially still be of interest even after many collisions because they they do not require a large kinetic energy to be of use for the plasma. As shown in section 6.1, cold electrons and ions are of use for the plasma as long as they are confined, no matter at which kinetic energy. However we will see that loss cone scattering limits their confinement lifetime to a single 90° collision or even less.

The lifetime of a particle confined in an mininum-B ECRIS confinement is given by [11]:

$$\tau = \tau_{90°} \, \log\left(\frac{B_{max}}{B_{min}}\right) \quad (3.16)$$

τ is the confinement time of the particle, $\tau_{90°}$ the collision time for a 90° scattering and B_{max}/B_{min} equals the mirror ratio of the axial magnetic mirror. $\log(B_{max}/B_{min})$ results in 0.3. This again shows, the collective confinement performance of a magnetic mirror significantly deviates from single particle motion, presented in section 3.2.1. Hence τ results in $\tau_{90°} \cdot 0.3$ which is even less than the collision time for a 90° scattering. Diffusion across magnetic field lines is always coupled to several 90° scattering collisions and is hence not of interest because it is dominated by the loss cone limiting particle life time τ.

With SWISSCASE parameters of CO$_2$ operation we can estimate $\tau_{90°}$ and τ by using table (3.1) and:

$$\tau_{90°} = \nu_{90°}^{-1} = \frac{1}{\sigma \, n \, v} = 7.36 \cdot 10^{-3} s \;\Rightarrow\; \tau = 2.21 \cdot 10^{-3} s \quad (3.17)$$

$\sigma = 2 \cdot 4.2 \cdot 10^{-20}$ m^2 is the dominating neutral neutral collision cross section, $n = 4.8 \cdot 10^{18}$ 1/m^3 the neutral density and $v = 336.8$ m/s the mean neutral velocity at 300 K. $\tau = 2.21 \cdot 10^{-3}$ s is the time singly charged ions are given to produce an ion charge state distribution resulting in the measured ion beam spectra presented in Chapter 7.

3.4 Plasma potential

In this section, the negative ECR plasma potential is detailed and the charge imbalance calculated which is necessary to establish this potential. The ECR plasma is charged negatively because the ECR electrons are better confined by the minimum-B structure than the ions due to pronounced anisotropic velocity distribution (see Section 3.2.1) of the ECR electrons. The negative charge of the ECR plasma significantly improves the confinement performance of the ions [11].

The plasma potential of minus 3 V (K.S. Golovanivsky and G. Melin, 1992 [14]) is caused by a slight majority of electrons over the ions. Knowing the size of the ECR zone and simplifying the situation by assuming a homogeneous charge distribution inside the ECR zone (see figure 3.9), we can calculate the charge majority by Maxwells first equation [20]:

$$\int_{\partial V} E \, dA = \frac{1}{\epsilon_0} \int_V \rho \, dV \qquad (3.18)$$

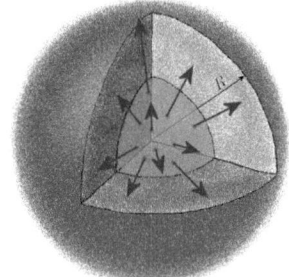

Figure 3.9: The ECR zone is approximated as a homogeneous cloud of surplus electrons. The space charge of the electrons creates an electric field. A standard integration from the cloud center to its outer radius yields the potential difference created by the electric field.

The electric field resulting from (3.18) is given by:

$$E(r) = \frac{r\rho}{3\epsilon_0} \qquad (3.19)$$

Integrating (3.19) results in a potential difference of:

$$\Delta U = \frac{\rho}{6\epsilon_0} R^2 \qquad (3.20)$$

With a potential difference of $\Delta U = 3$ V [14, 17] we can solve for ρ to obtain a surplus electron density of $2.76 \cdot 10^{13}$ $1/m^3$. Considering the total plasma electron density of $1.46 \cdot 10^{18}$ $1/m^3$, the surplus electron density amounts to $1.89 \cdot 10^{-5}$ of the total electron density.

The potential created by the slight surplus density of the ECR electrons attracts positively charged ions toward the ECR zone and repels plasma electrons. With no magnetic field present this leads to straight trajectories for each particle toward or away from the ECR zone depending on its charge. This suggests an additional plasma electron loss. However in SWISSCASE and any other ECR ion source there is a strong magnetic field present leading to $E \times B$ drift [7, 11, 26, 16]. Hence the charged particles would feature a circumferential cycloid motion rather than a radial straight one. In real minimum-B ECRIS the magnetic field is not a simple magnetic mirror but a superposition of a magnetic mirror in axial direction and four to 24 (in hexapole sources: 6) radial magnetic mirrors. This leads to a highly complex spatial magnetic field distribution (see figure 3.10).

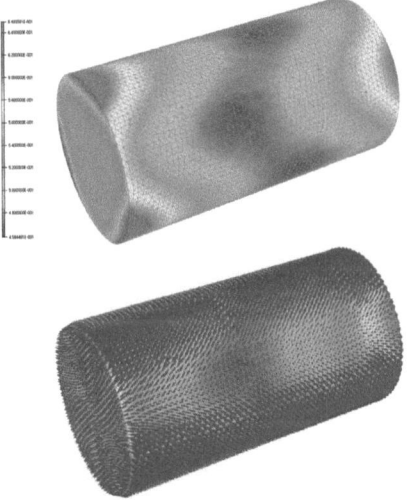
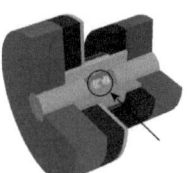

Figure 3.10: Cylinder cut of the SWISSCASE magnetic FEM model the ECR zone. The cylinder has a diameter of 14 mm and a length of 30 mm. The top picture features the magnetic field distribution on the surface. The lower picture shows the highly complex magnetic field vectors distribution. The color bar starts at 0.46 T and stops at 0.65 T. To the right: 3D overview.

Figure 3.10 shows the complex magnetic field resulting from the superposition of the axial magnetic mirror and the radial Halbach hexapole in SWISSCASE. The spatial magnetic field distribution of SWISSCASE and any other minimum-B ECRIS renders the analytical investigation of single particle motions in these confinements futile. However numerical trajectory integration is accessible with the magnetic field simulation (see chapter 6).

3.5 Plasma density

During an ECR process the ECR plasma approaches a quasi stable dynamic equilibrium with a plasma density and a charge state ion distribution. Despite the many operational parameters that can be chosen, the maximal plasma density is limited by the microwave frequency and the coupling mode. In SWISS-CASE the exact microwave coupling is highly complex and is only accessible by numerical simulations of electric and magnetic field propagation. This is caused by the plasma cavity dimensions creating a near-field situation rather than a far-field situation of electro magnetic wave propagation. In addition the ECR zone of SWISSCASE has a length of 13 mm, only a fraction of the wavelength of the incident microwave denying the assumption of a long plasma column.

However first order approximations lead to useful plasma densities supported and made plausible by feed gas pressure, mean free path lengths (Section 3.2.2) and power balance considerations (Chapter 3). This first approximation assumes a right hand circularly polarized, an R-mode, coupling between incident microwave and ECR plasma [7, 11].

An R-mode of frequency f_{mw} can only propagate inside a plasma if the plasma electron frequency f_{pe}, called the plasma cut off frequency, is lower than than f_{mw}. Otherwise the incident microwave is reflect and a coupling is prohibited. We shall refer to this property as plasma opacity. Applying energy conservation we can distinguish two basic situations:

- $f_{mw} > f_{pe}$. The ECR electrons can be heated and the plasma density is allowed to increase. The plasma is under-dense.

- $f_{mw} < f_{pe}$. The ECR plasma is opaque with respect to the incident microwaves and the ECR electrons are not heated. The plasma is over-dense and there is cutoff. Hence, it is losing thermal energy and is cooling. The plasma density decreases.

This leads to a bistable situation because both, under-dense plasmas and over-dense plasmas approach the critical density. Assuming this bistable equilibrium we can set: $f_{mw} = f_{pe}$. The plasma electron density f_{pe} is given by [7]:

$$f_{pe} = \frac{1}{2\pi} \sqrt{\frac{n_e\, e^2}{\epsilon_0\, m_e}} \qquad (3.21)$$

n_e is the plasma electron density, e the charge of the electron and m_e its mass. Rearranging (3.22) and combining it with $f_{mw} = f_{pe}$ leads to:

$$n_e = \frac{4\pi^2 \epsilon_0\, m_e}{e^2}\, f_{mw}^2 \qquad (3.22)$$

Hence the plasma density scales with the square of the microwave frequency f_{mw}. A higher microwave frequency allows a much higher plasma electron density and provides a better environment for the production and confinement of highly charged ions. In addition a higher plasma electron density allows to design the ECR zone more compact because the same number of plasma electrons fit into a tighter volume. This again significantly increases power efficiency due to a smaller ECR zone surface and related particle losses. In addition smaller space requirements and reduced magnet confinement cost can be expected.

For SWISSCASE n_e is $1.46 \cdot 10^{18}$ $1/m^3$. This is the maximal plasma electron density limited by the opacity of the ECR plasma with respect to the incident microwave. If not chosen optimal different operational parameters, such as power input, feed gas pressure or microwave impedance matching, result in a significant reduction of the given electron density and the resulting ion beam performance.

3.6 Comparison of SWISSCASE and MEFISTO

Despite sharing the same principle of operation electron cyclotron resonance ion sources can be different in their design and operational parameters. We compare MEFISTO and SWISSCASE, both designed, built and operated at the University of Bern. Table 3.2 gives a summary of the essential confinement and the optimal plasma parameters.

Table 3.2 shows that the Larmor radii of hot electrons and for H^+ ions are similar in SWISSCASE and MEFISTO respectively. This suggests that hot electrons and ions with a low m/q ratio feature similar confinement performance. However the large anisotropic velocity distribution and the low collision cross section of hot electrons results in a much better hot electron confinement and less diffusion across magnetic field lines. The negative plasma potential however favors the ions, due to their positive charge, rather than the hot electrons, somewhat compensating for the weaker magnetic confinement of the ions.

In addition SWISSCASE and MEFISTO feature similar power densities with respect to the ECR zone volume and its surface. SWISSCASE features better ion beam performance with higher input frequency and lower input power despite its far smaller ECR zone volume. The higher microwave frequency allows a square times higher plasma electron density. This results in an ionization fraction of 30.25 % for SWISSCASE and 1.53 % for MEFISTO. The total electron count inside the ECR zone is similar for both SWISSCASE and MEFISTO. However the ratio N_e/A of total electrons and ECR zone surface is in favor for SWISSCASE by more than an order of magnitude. This suggests that SWISSCASE suffers less from particle loss through the ECR surface and radiates less cyclotron emission power at the same plasma parameters than MEFISTO. This results in a better particle confinement and a hotter plasma with ECR electrons of higher energies. The latter is supported by a Bremsstrahlung measurement presented in Chapter 8.

3.7 Conclusion

The numerical particle trajectory integration to simulat the ECR acceleration process showed that:

- the randomized initial phase of the particle gyration with respect to the incident microwave influences the early part of the acceleration process but leads to very similar energy behavior once larger electron energies are achieved.

ECR zone	SWISSCASE (D 12, L 13)	MEFISTO (D 20, L 37)
Vol	0.980 cm^3	8.61 cm^3
A	4.78 cm^2	21.41 cm^2
P_{max} @ f_{mw}	95 W	300 W
P_{max}/Vol	9.692·10^7 W/m^3	3.481·10^7 W/m^3
P_{max}/A	1.989·10^5 W/m^2	1.401·10^5 W/m^2
f_{mw}	10.88 GHz	2.45 GHz
λ_{mw}	27.55 mm	122.36 mm
B_{ECR}	388 mT	87.5 mT
n_e	1.46·10^{18} 1/m^3	7.41·10^{16} 1/m^3
N_e	1.431·10^{12}	6.380·10^{11}
N_e/A	2.99·10^{15} 1/m^2	2.980·10^{14} 1/m^2
T_{ehot}	10 keV	1.7 keV [17]
T_{ecold}	2 eV [11]	2 eV [11]
κ_i	30 %	1.53 % [17]
r_{Lehot}	8.68·10^{-4} m	1.59·10^{-3} m
r_{Lecold}	1.23·10^{-5} m	5.45·10^{-5} m
r_{LH+}	5.26·10^{-4} m	2.34·10^{-3} m
r_{LC+}	1.82·10^{-3} m	8.09·10^{-3} m
r_m	2.39, 1.95	4

Table 3.2: Summary of design and optimal plasma operation parameters for SWISSCASE and MEFISO. ECR zone dimensions in mm, Vol and A: ECR zone volume and surface assuming a rotational elliptical zone shape, f_{mw}: microwave frequency, P_{max} @ f_{mw}: maximal microwave input power at specified microwave frequency, P_{max}/Vol: specific microwave power per ECR zone volume, λ_{mw}: wavelength of microwave, B_{ECR} ECR qualified magnetic field, n_e: plasma electron density limited by plasma opacity, N_e: total electrons inside the ECR zone, T_{ehot}, T_{ecold}: hot and cold electron temperature, κ_i: ionization fraction, r_{Lehot}, r_{Lecold} Larmor radii for hot and cold electrons, r_{LH+}, r_{LC+} Larmor radii for H^+ and C^+ ions, r_m: mirror ratio.

- the necessary acceleration path length is 0.27 m to 0.372 m and the electrons complete 108 to 110 gyro revolution, depending on the initial phase shift.

- the simulation is supported by the critical mean free path of 0.49 m, a maximum feed gas pressure equivalent above which SWISSCASE ceases to output multiple charged ions. In addition the simulation is also supported by the experimentally obtained optimal mean free path of 4.9 m which enables ECR electrons to be accelerated and scattered about ten times before hitting a neutral particle.

Hexapole confined ECR plasmas can be approached analytically with simplified single particle motion and simple magnetic mirrors. However this approach is limited to the very basic understanding of ECR ion sources. More insight is available by the study of collective plasma behavior including collision mechanisms, diffusion and numerical field and trajectory simulations. Collision mechanisms are useful for the study of the ion distribution because they yield collision frequencies and mean free paths. Diffusion processes in ECR plasmas however are still widely unexplored and useful models are not yet available. On the other side present diffusion theory can be used to estimate the collision number a particle undergoes before hitting the plasma chamber wall further detailed in chapter 6.

A comparison between SWISSCASE and MEFISTO suggests a smaller particle loss via the ECR zone surface resulting in a better confinement and lower radiation loss for SWISSCASE. SWISSCASE also features a larger critical plasma density and power-volume density than MEFISTO, resulting in higher ion beam currents of highly charged ions.

Chapter 4

Realization of source elements

This chapter is structured corresponding to the different components of the ECR ion source. Figure 4.1 gives a 3D overview, Figures 4.2 and Figure 4.3 show a top and a side view of the realized components of SWISSCASE.

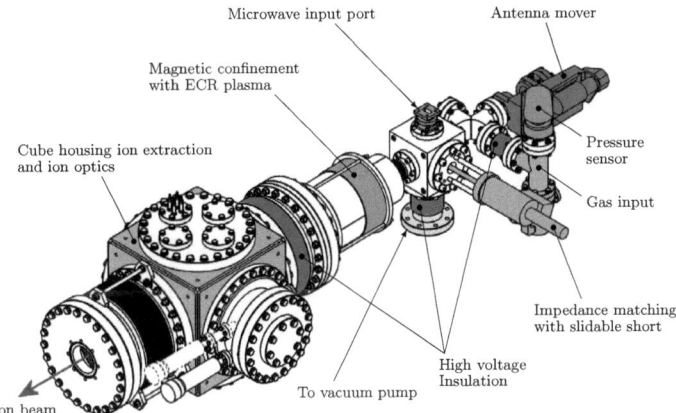

Figure 4.1: A 3D overview of SWISSCASE realized within this thesis.

In Section 4.1 the magnetic confinement responsible for an efficient electron cyclotron resonance and an successful ionization process is presented. Section 4.2 presents the microwave system used for the electron heating via ECR effect and the necessary subcomponents. In Section 4.3 the newly designed extraction system is introduced and effects of different plasma operation on the puller electrode explained. Section 4.5 describes the newly designed ultra high vacuum (UHV) facility housing the plasma chamber, the extraction and the ion optics.

36 Chapter: Realization of source elements

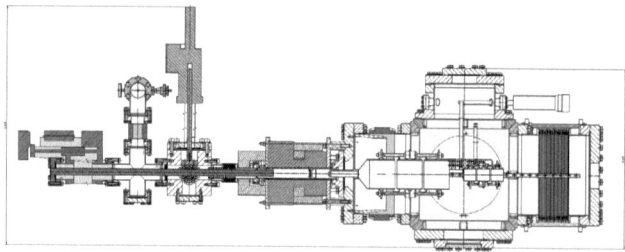

Figure 4.2: Top view section cut through SWISSCASE. Color coding is identical to Figure 4.1. The width indicated on the left hand side measures 599 mm, the width to the right hand side measures 450 mm.

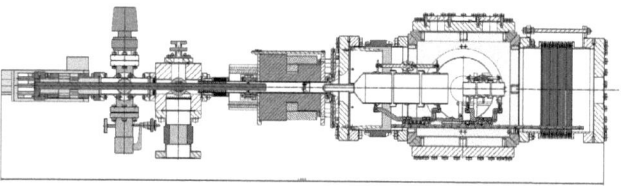

Figure 4.3: Side view section cut through SWISSCASE. The length of the facility measures 1499 mm.

Section 4.6 presents the high voltage setup necessary for the ECR plasma ion extraction and the ion optics. The chapter concludes with Section 4.7, which gives a detailed calculation of the necessary parameters of the mass separation magnet used for spectral analysis of the extracted ion beam.

This chapter is a description of the mechanical and the electrical setup of the ion source SWISSCASE and justifies some of the purchase criteria of the microwave generator, the mass separation magnet. For a detailed discussion of the ion source plasma parameters please refer to Chapter 7. For details about magnetic field and electron distribution inside SWISSCASE please refer to Chapter 6.

4.1 The magnetic confinement

4.1.1 Overview

The magnetic confinement assures a successful ECR process to heat plasma electrons to energies large enough for electron impact ionization. Neutral atoms and molecules are ionized by electron neutral collisions and finally establish a quasi stable plasma with electrons and ions. Since ionization to high charge states is a step by step process [16] the ions have to be confined for a time long enough to undergo multiple collisions to obtain multiple charge states. For the detailed magnetic field distribution, both measured and simulated, please refer to Chapter 6. The axial magnetic confinement is provided by four ring magnets arranged around a Halbach hexapole. Figures 4.4 (a) and (b) show section cuts featuring a side view and a front view of the magnetic confinement used in SWISSCASE for magnetic confinement and for support of the ECR process.

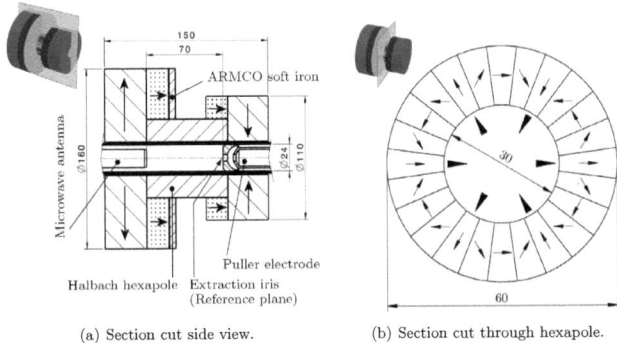

(a) Section cut side view.   (b) Section cut through hexapole.

Figure 4.4: (a) Side and (b) front section cuts of the magnet assembly used in SWISSCASE. Arrows indicate the direction of magnetization of the permanent magnetic material. (a) shows the magnetization of the ring magnets together with the plasma chamber, the microwave antenna and the ion extraction system. (b) reveals the characteristic alternating orientation of 24 permanent magnet sectors. Bold arrows indicate the resulting main magnetic poles inside the hexapole. Dimensions are in mm.

4.1.2 Manufacturing process

The magnetic confinement system is established by a sophisticated arrangement of permanent magnets. It was manufactured by Vacuumschmelze Hanau based on a design by Salzborn, Trassl and Broetz et al. [6, 36] at the University of Giessen. Vacuumschmelze Hanau has already manufactured several permanent magnet confinement systems for the University of Giessen and they alone have the know-how of bonding, agglomeration (sintering) and overall manufacturing of such an assembly. The design already proved to be reliable and capable of a high performance confinement resulting in the successful production of high intensity beams of highly charged ions at the ion beam facility located at the Beam Center of the Giessen University [6, 36]. Despite sharing the same geometry with the Giessen ECR ion source the SWISSCASE ion source is manufactured of new VACODYM 677 HR rather than VACODYM 400 HR and VACODYM 335 HR. VACODYM 677 HR promises a higher residual magnetic induction. This change toward a material with a higher residual induction modifies the configuration of the magnetic field and increases the field density in the minimum region while keeping the geometry constant. This in turn changes the location of the ECR zone.

4.1.3 Assembly process

To ease the assembly process of the magnetic confinement system an assembly facility has been constructed. The assembly facility enabled a safe and reliable assembly process. However once assembled the confinement system is not supposed to be disassembled. An attempt to disassemble the confinement system may result is the destruction of one or several permanent magnets because the attractive forces created by the magnetic field locally exceed the shear and tensile strength of the material.

The magnetic confinement system is mechanically instable. All four disc shaped magnets tend to slip apart side ways. It is therefore of vital importance to always fill out the inner hollow cylinder shaped volume with a solid material such as a PVC tube, a broom stick with the correct diameter or any other non ferromagnetic material.

The Halbach hexapole is unstable in itself too. It is bonded by high performance glue by Vacuumschmelze Hanau. This bonding can fail (Salzborn et al., 2002). The resulting explosive disintegration of the Halbach hexapole magnet can seriously harm human beings and nearby devices. To prevent such an event an aluminum bracelet encapsulating the Halbach hexapole magnet has been implemented.

4.1.4 Optimal setup

Unsuccessful attempts to ignite the ECR plasma operated with a CO_2 working gas were the result of experiments with the future SWISSCASE confinement performed by Bodendorfer and Trassl at the University of Giessen from July 2005 to August 2005 using the unaltered magnetic confinement configuration and a magnetron limited to 10.5 GHz. This was due to the unexpected high magnetic field inside the hexapole resulting from the new VACODYM 677 HR material. Adding soft iron plates at both inner shoulders of the confinement magnet bypassed some of the magnetic field and lowered the central magnetic field where the ECR zone is located to a value which fulfilled the ECR condition with 10.5 GHz. Subsequent experiments by Bodendorfer et al. in Giessen during August 2005 showed that the ECR plasma would ignite with various combinations of 2 mm and 4 mm ARMCO soft iron plates and the given magnetron. However optimal ion beam extraction performance for Ar^{8+} was achieved with two 4 mm soft iron plates each one attached to one magnet shoulder.

Experiments with the final setup performed at the University of Bern during February 2008 showed that, thanks to the higher available frequency of the solid state microwave generator, higher values for the resonant magnetic field were accessible and that another optimum with respect to ion output was found using only one 4 mm ARMCO soft iron plate attached next to the microwave injection (see Figure 4.4a). This new optimal setup operated at 10.88 GHz and one soft iron plate is superior to the setup operated at 10.5 GHz with two soft iron plates in so far that it produces ion currents increased by 23% to 29% for Ar^+ to Ar^{12+}. Figure 4.5 gives the measured magnetic field of SWISSCASE along

Chapter: Realization of source elements

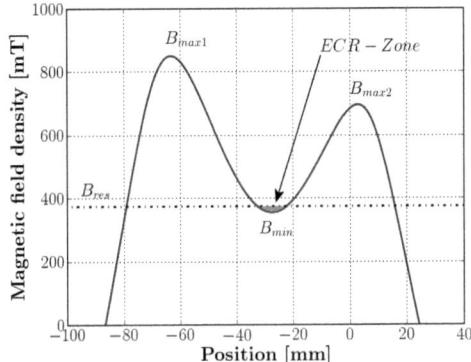

Figure 4.5: Magnetic field along z-axis of SWISSCASE. One 4 mm Armco soft iron plate is attached to the magnet shoulder according to Figure 4.4. Z-coordinate origin is at the reference plane defined in Figure 4.4a.

the z-axis in the final state with one single soft iron plate and indicates B_{res} as the magnetic field value fulfilling the ECR condition.

This setup turned out to be close to the one used by Broetz at al. in 2000. Table 4.1 compares the high B-field confinement setup used by Broetz at al. in 2000 with VACODYM 335/400 HR and the SWISSCASE confinement using VACODYM 677 HR and one 4 mm soft iron plate.

Facility	High-B field	SWISSCASE
Material	VACODYM 335/400 HR	VACODYM 677 HR
Magnetic bypass	none	4mm ARMCO soft iron
Operation frequency	$9.97\ GHz$	$10.88\ GHz$
Field maximum 1	$780.21mT^* \pm 18mT$	$850.5mT \pm 1mT$
Field maximum 2	$673.82mT^* \pm 18mT$	$696mT \pm 1mT$
Field minimum	$354.64mT^{**} \pm 0.2mT$	$356.3mT \pm 1mT$
Mirror ratio 1	2.2 ± 0.05	$2.387 \pm 4 \cdot 10^{-4}$
Mirror ratio 2	1.9 ± 0.05	$1.953 \pm 9 \cdot 10^{-4}$

Table 4.1: Comparison of High-B-field confinement from Broetz at al. (2000) with SWISSCASE. ** calculated by Eq. 3.1 (ECR eq.) using $f_{ECR} = 9.97GHz$ from [6]. * calculated by using mirror ratio 1 and 2.

Table 4.1 shows the mirror ratios of SWISSCASE and the ECR ion source by Broetz at al. in 2000 differ by 8.5 % for the high and by 2.8 % for the lower mirror ratio. This suggests [11] that SWISSCASE is able to perform better due to its increased mirror ratio and consequently smaller loss cone (see Chapter 3) while maintaining a similar field minimum compared to Broetz at al. in 2000.

4.1.5 Temperature regime of operation

VACODYM 677 HR was a new material at the time the magnetic confinement of SWISSCASE has been manufactured. Both its high residual induction of 1.18 T and coercive magnetic field strength of 2466.9 kA/m (corresponding vacuum magnetic flux density: 3.0999 T) promised a very good confinement performance. Figure 4.6 shows the demagnetization curve of VACODYM 677 HR.

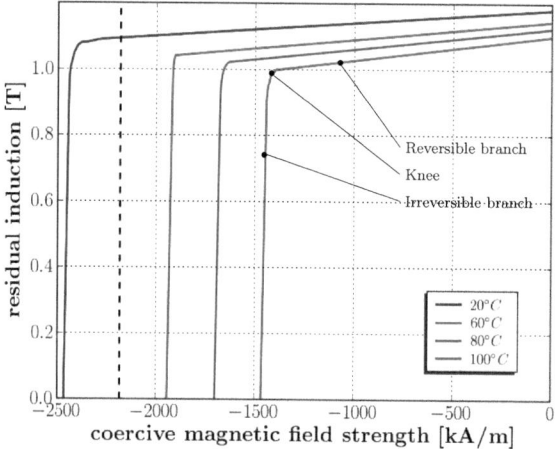

Figure 4.6: Demagnetization curves of VACODYM 677 HR for different temperatures. The dashed vertical line indicates a coercive magnetic field strength of $2179 kA/m$ representing the maximal field strength given by the FEM model. Data by Vacuumschmelze Hanau.

If the temperature of the magnetic material rises, the magnetic field density decreases. This temperature dependent demagnetization is reversible if the coercive field strength stays within the upper shoulder, the reversible branch, of the demagnetization curve (see Figure 4.6). If the coercive field strength further decreases below the knee point, the magnetic material enters the irreversible demagnetization branch and permanent demagnetization occurs. This effect of permanent demagnetization spreads out from the locations of highest field densities to locations with lower field densities if the temperature is further increased. It is therefore important to keep the temperature of the magnet assembly below the critical temperature where demagnetization starts at the locations of highest field densities.

To find this critical temperature we first evaluate the highest field density in the magnet assembly using the finite element model presented in Section 6. Figure 4.7 shows a 3D simulation of the confinement featuring the zones of highest magnetic field.

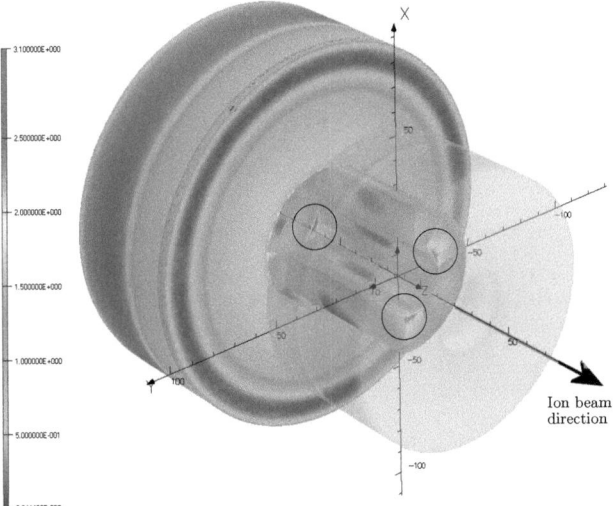

Figure 4.7: 3D overview of the SWISSCASE magnetic field simulation. Both small ring magnets and halve of the hexapole are shaded transparent to enable the view to the critical locations where the magnetic flux density reaches its maximum value. Circles indicate three positions of maximal flux density. Due to symmetry there are three more positions which are not visible in this view.

The FEM model gives a maximal magnetic field strength of $2.179 \cdot 10^6\ A/m$. A linear interpolation between the two demagnetization curves of 20 C° and 60 C° results in a critical temperature of 37.2 C°. However as the FEM model shows the locations of such high field strengths are very limited in spatial extent. A local permanent demagnetization is not supposed to destroy the whole magnetic confinement but could change it to such an extent that the ECR zone is no longer in an optimal position. This in turn could significantly decrease the ion beam performance of SWISSCASE. To prevent such a scenario safety measures have been implemented to keep the magnet temperature below the critical temperature.

During assembly of the different permanent magnets into the definite confinement system large forces have to be overcome. The repulsive and attractive forces and especially their inverse square dependence on the distance of two magnets easily exceed the abilities of a strong human being. Chipped, cut or crashed fingers are the result of handling the separate magnets without the necessary caution.

4.2 The microwave system

The microwave system is responsible for the ECR heating of the plasma electrons. Different components are implemented to satisfy the needs of a proper coupling between the microwave generator and the plasma. In such an optimized configuration the generator represents an ideal microwave power source and the plasma represents an ideal microwave power sink. However neither the real plasma nor the real microwave power generator act as a pure sink or a pure source. The nature of the plasma is such that in many operation conditions the majority of the incident microwave power is reflected back toward the microwave power generator. To prevent the reflected microwave power causing a malfunction in the microwave power generator a circulator has been implemented further discussed in Subsection 4.2.3.

Due to the optimal operation frequency of 10.88 GHz and the corresponding wavelength of 27.55 mm the impedance of the system significantly depends on the system geometry, the plasma shape, its condition and location. During ECR operation the complete system undergoes thermal expansion caused by the conversion of microwave power to heat in the plasma chamber and caused by changes in the room temperature. The plasma further changes both spatial location and parameters in Smith space [16] depending on operational condition. During experimental verification also different frequencies had to be tested. All these varying microwave conditions called for a system which is able to compensate the effects and establish proper coupling between microwave power generator and the plasma. This system is called an impedance matching system and will be discussed in Subsection 4.2.4.

Figure 4.8 gives a schematic overview of the microwave system with the most relevant components.

4.2.1 The microwave generator

The microwave generator serves as power source to increase the plasma electron's kinetic energy via the ECR process described in Chapter 3. First we show the criteria for the selection of a suitable microwave generator, followed by detailed information about its characterization and its implementation into the wave guide system. Finally we will have a closer look into the microwave coupling structure, which insures coupling between the microwave generator, its protection from reflected radiation and impedance matching between microwave source and load.

The criteria for a suitable microwave generator are given by the application. The magnetic confinement features a minimal magnetic flux density of 356.3 mT requiring a minimal ECR feed frequency of 10.02 GHz. Because of the need for experimental verification of the optimal operation frequency and therefore the optimal localization of the ECR zone the microwave generator has to deliver a variable frequency rather than a fixed one. On the other hand available wave guides in the common X-band are suitable for a frequency between 8.2 GHz and 12.4 GHz. This gives an upper limit for the frequency range of the microwave generator. While a higher output frequency than 12.4 GHz would not damage the system it would undergo severe attenuation by the wave guide system, the majority of the microwave energy would be converted into heat before reaching the plasma. Testing the magnetic confinement at the University of Giessen

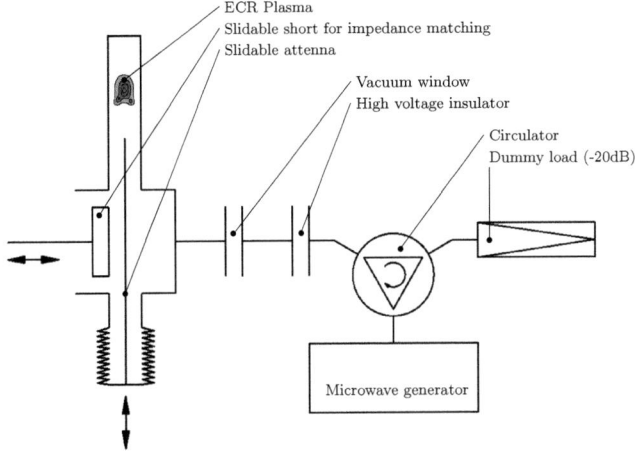

Figure 4.8: Overview of the microwave system with the most relevant components labeled.

showed a significant decrease in ion output for operation frequencies below 10 GHz. The same test suggested more ion output for frequencies above 10.5 GHz. This frequency range could not have been investigated at the University of Giessen because the available magnetron microwave generator was limited to 10.5 GHz.

Experiments at the University of Giessen [36] have also shown that the optimal input power is around 80 W for this kind of system. At the point of choice this has been the only available data for this magnetic setup. Later experimental verification at the University of Bern including the new purchased 100 W microwave generator suggested the possibility to further increase the ion performance by exceeding the input power of 100 W (see Chapter 7).

For ease of operation both output power and output frequency have to be controlled by two front panel knobs. A power meter has to indicate the effective output power and a frequency clock has to indicate the chosen output frequency. In addition to this both output power and output frequency have to be accessible by an electrical terminal at the back of the device for the implementation into a computer controlled measuring and control system.

For the protection of the magnetic confinement from thermal overload the back terminal has to facilitate a shut down connector, which can be integrated into a temperature control circuit. This allows to switch off the microwave generator as the only power source and therefore heat input, of the plasma to let the plasma chamber and the magnet assembly cool down.

SWISSCASE will finally be operated on a high voltage terminal. The microwave generator primary power will have to be supplied by a high voltage isolation transformer. Hence the primary power of the microwave generator was also a purchase criterium to save on the necessary installed electrical feed power of the high voltage isolation transformer. A single phase power supply is strongly favored rather than three phase again in favor of an economic high voltage terminal power supply.

Table 4.2 gives a summary of the quantitative purchase criteria of the microwave generator.

Output power	0 – 80 W
Output frequency	10 – 11 GHz
Primary power	< 750 W
	single phase
Cost	< 60'000 CHFr.

Table 4.2: Quantitative requirements for the purchase of the microwave generator.

4.2.2 Choice and characterization of the microwave generator

A total of eight quotes from four different manufacturers have been received. Table 4.3 shows the three most attractive options of the microwave offers.

	WaveLab Eng.	Microwave Power	Inwave
Frequency range [GHz]	9.5 – 11.0	9.0 – 11.5	9.0 – 11.0
Power output [W]	100	100	100
Principle of operation	TWT	SS	TWT
Amplifier manufacturer	Quarterwave	Omniyig	Quarterwave
Price Excl. VAT [CHFr.]	115'600	57'000	77'500

Table 4.3: Summary of purchase options for the microwave generator. TWT: Traveling wave tube amplifier. SS: Solid state amplifier.

From Table 4.3 it is clear that the solid state microwave power amplifier is the only purchase option that fits within the budget of 60'000.- CHFr. The specification fulfill all the requirements. It has therefore been decided that the microwave generator of choice is a solid state RF generator from Microwave Power Inc. [28]. The output frequency can be varied from 9 to 11.5 GHz. The generator output is specified to be 100 W or more between 10 and 11.5 GHz. Table 4.4 gives a summary of the specifications provided by Microwave Power Inc.

However our own measurements resulted in significantly different power output data than specified by Microwave Power Inc. We installed a measurement system consisting of the microwave generator, a directional coupler with a coupling of -20 dB, two subsequent attenuators of -10 dB each and a LabView measurement system (see schematic in Figure 4.9).

All components of the measurement system between the microwave generator and the power sensor feature a frequency dependent attenuation of the

Output power	0 – 100 W
Output frequency	9 – 11.5 GHz
Input power	600 W single phase
Principle of operation	solid state
	Yig tuned oscillator,
	open frequency loop, open power loop,
	6x MPI power modules at 41 dBm each

Table 4.4: Summary of specification of the microwave generator.

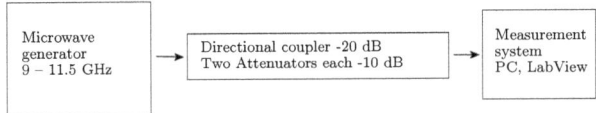

Figure 4.9: Composition of the power measurement system. The microwave generator is coupled with the computer measurement system with a total attenuation of -40dB.

microwave signal. To compensate for this frequency dependent attenuation a separate calibration has been performed to evaluate the exact frequency dependent transmission characteristics of the connecting parts. Figure 4.10 shows the frequency dependent output power corrected by the attenuation of the measurement system.

The measured power output significantly deviates from the microwave generator performance as specified by microwave power inc. Despite this deviation from the specified power output we accepted the generator due to timetable considerations (delivery time was close to one year) and because the measured power output is sufficient in the important frequency band of 10 to 11 GHz to operate the ECR ion source based on data by Broetz et al. [6]. All subsequent ion beam performance measurements are based on this varying power output. Therefore care has to be taken in comparing the ion beam performance of different operation frequencies especially at a constant microwave power setting.

4.2.3 Wave transport system and circulator

The wave transport system transmits the microwave power from the microwave generator to the ECR plasma. It has to fulfill different criteria presented in this Section.

Back traveling microwaves entering the microwave generator can destroy the generator. The reflected power tolerance is 25W. The generator features an internal load to absorb this specified reflected power. The high dynamics of the plasma deny a precise forecast of the ratio of reflected to input power. Hence we have to assume that in certain conditions almost all the input power of 100W will be reflected. In such a case the reflected power tolerance of the internal

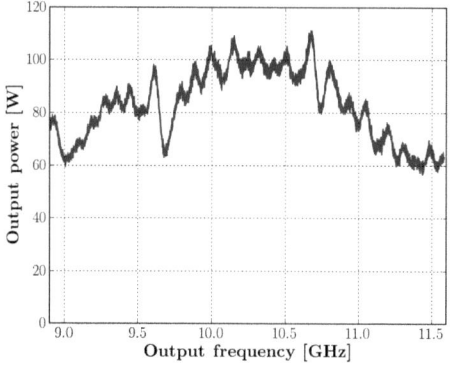

Figure 4.10: Measured power output of the microwave generator from Microwave Power Inc. The power output varies between 62W and 109W. In the frequency range between 10 and 11 GHz the output power is sufficient.

load would be exceeded. Therefore, an additional external circulator is installed in the transmission line to prevent microwaves reflected from the plasma or the plasma container to enter the microwave generator. The circulator is acting similar to a diode (see Figure 4.11. It can distinguish between forward and backward traveling microwaves and separates them.

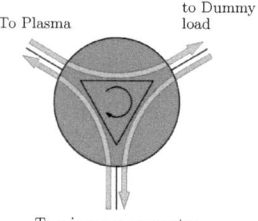

Figure 4.11: Schematic of a ferrit core circulator. A circulator 'circulates' the microwaves in a defined orientation (here clockwise) and distinguishes between forward and backward traveling microwaves. It serves as protection of the microwave generator from back traveling microwaves reflected from the plasma or the plasma container and the wave guide system.

To insulate the microwave generator from the high voltage (see Section 4.6 a high voltage insulator microwave window is implemented in the wave guide system. The high voltage insulator microwave window consists of 1 mm thick PTFE teflon plate and a composite structure insuring precise alignment of the transmitting and the receiving wave guide connectors.

The wave guide system is operated under atmospheric pressure. Hence it has to be separated from the UHV system by a UHV microwave window. This UHV microwave windows is built in copper and tested for UHV capability. The vacuum seal and the microwave transmission is defined by a 2.3 mm ceramics plate soldered to the copper wave guide stubs featuring standard X-Band

Chapter: Realization of source elements 47

connectors on both sides.

4.2.4 Impedance matching

Since the plasma represents an active element in the microwave system with an impedance that is not constant with respect to time nor to different operation conditions an impedance matching system was implemented. To meet the needs of coupling the microwave generator via wave guide system, vacuum window and high voltage window with the plasma a slidable short is attached to the coupling cube. The remote controlled short can be moved perpendicular to the optical axis. The remotely controlled access allows the operation of the short during active extraction when the ECR ion source is connected to high voltage and manual access is prohibited. Figure 4.12 shows a section cut of the implemented short serving as impedance matching system.

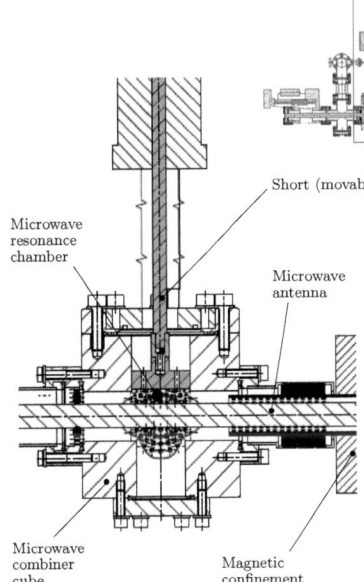

Figure 4.12: Section cut of the short implemented for impedance matching of the microwave power generator and the ECR plasma. The short can be moved parallel to its own axis for impedance matching by changing the interior geometry of the resonance chamber and the distance to the microwave antenna. Note: the microwave input port is oriented perpendicular to the plane and not shown in this figure.

4.3 Ion optics

The ion optics has to extract the ion beam from the plasma and transport it from the extraction to the target where the beam is used. On the way from the extraction to the target the ion beam has to pass through different vacuum vessels and a ninety degree mass separation magnet, which separates the ion beam into a mass per charge resolved ion band for spectral analysis. In the following section the first stage of the ion optics will be presented, the extraction process of the ions.

4.3.1 Extraction

The extraction consists of an extraction aperture and a puller electrode and extracts the ions from the ECR plasma to form an ion beam. Figure 4.13 shows the principle of ion extraction by high voltage.

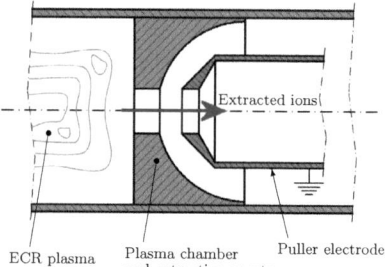

Figure 4.13: Section cut of extraction setup. The puller electrode colored in blue is grounded. The plasma chamber colored orange is connected to high voltage. The applied high voltage defines the kinetic energy of the extracted ions.

For the start up of the new ion source the extraction mechanism from the University of Giessen has been used to exclude as many unknown effects as possible during first light of the new ion source in Bern. As second step the extraction mechanism has been redesigned resulting in a significant increase in mechanical precision and far better vacuum performance. These improvements allow a better access of the different extraction parameter regimes and significantly improved reproducability of the chosen operational parameters. Figure 4.14 shows the realization of the newly designed extraction assembly.

During operation ions are extracted from the ECR plasma by the application of a high voltage potential drop from the plasma toward the puller electrode. Accelerated ECR plasma ions that are not properly focused hit the puller electrode rather than flying through its central bore. This leads to wear of the puller electrode depending on different plasma parameters, acceleration defining extraction voltage and operation time. Hence the puller electrode is a consumable and has to be replaced after a certain run time. Figure 4.15 shows a photograph of different puller electrodes each exposed for a different time of operation.

Beside the extraction potential and hence the kinetic energy of the sputtering ions also the plasma background gas significantly influences the wear of the puller electrode. Figure 4.15 shows the difference in sputtering progress for an Ar plasma and CO_2 and O_2 plasma after approximately the same duration of operation. The sputtering visible on the tip of the puller electrode used to extract oxygen ions is significantly further progressed than on the tip of the puller electrode used to extract Ar ions.

After colliding with the puller electrode Ar^{n+} ions get neutralized to Ar atoms which have an extremely low reaction probability. On the other hand O^{n+} ions colliding with the puller electrode get neutralized to O radicals for a

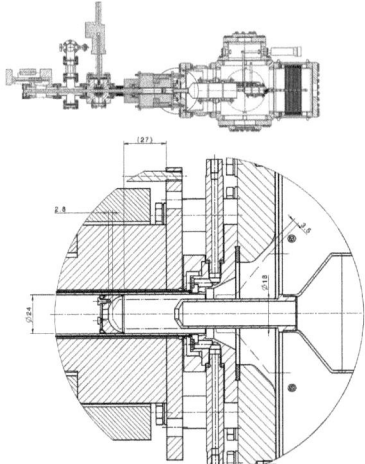

Figure 4.14: Realization of the newly designed extraction assembly. The puller electrode assembly can be slid toward the extraction aperture to a minimum distance of 2.8 mm. However with an applied extraction voltage of 10 kV this minimal approach results in electric arcing between the puller electrode and the extraction aperture. Hence the practical minimal distance is not limited by the mechanism.

Figure 4.15: Photographs of four different puller electrodes with the same original geometry and built from the same material of stainless steel. From left to right the puller electrodes have been in operation for 0h, $\sim 100\ h$ at Ar plasma extraction, $\sim 200\ h$ at Ar plasma extraction and $\sim 200\ h$ at CO_2 and O_2 plasma extraction. All puller electrodes were operated at 10 kV extraction potential. Note the difference in sputtering between the third and the fourth puller electrode despite the same exposure time but different plasma.

short time before further reacting with either other O radicals, gas molecules or metal lattice atoms from the puller electrode. It is clear that O radicals represent a far more aggressive chemical environment for the puller electrode

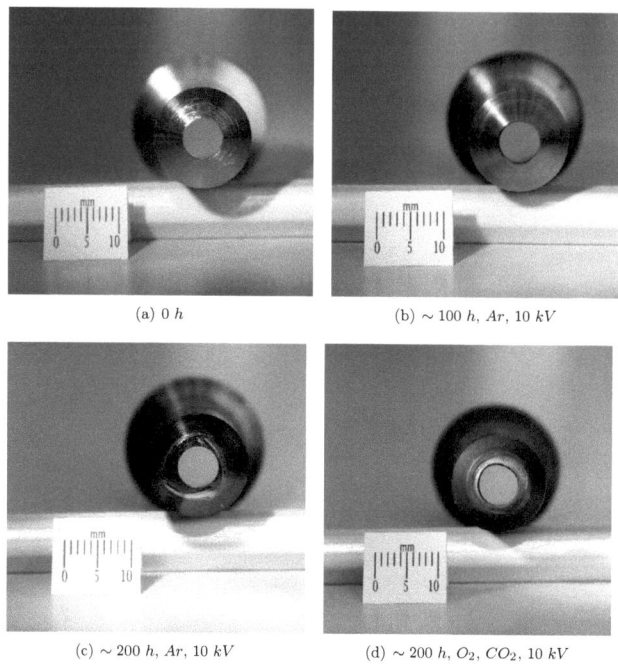

Figure 4.16: Photographs of the puller electrode tips exposed for different times and different plasma.

than Ar atoms. The presence of O radicals or neutral molecules leads to a partial oxidation of the surface layer of the puller electrode. This leads to embrittlement of the stainless steel material and makes the puller electrode tip more vulnerable for sputtering. Figures 4.16 a.) to d.) show photographs of the puller electrode tips from Figure 4.15.

4.3.2 Baffle

Because the plasma ions are extracted with an extraction voltage of 10 kV, the ions have a kinetic energy of 10 keV per charge unit. This energy is enough to remove atoms from the metal lattice of the puller electrode and sputtering occurs (see previous Section). The sputtered material travels for a certain mean free path before it either hits a neutral gas atom or molecule or it collides with a wall and gets neutralized. This way a surface coating with the material of the puller electrode can occur. Said material is conductive stainless steel and can destroy the insulation effect by bypassing an excessive leakage current through the surface coating.

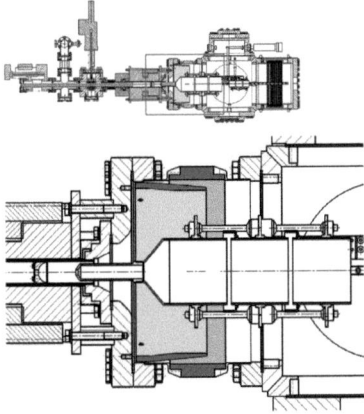

Figure 4.17: A section cut where the baffle (colored in red) is implemented in SWISSCASE. The baffle protects the high voltage insulation ceramics (colored in magenta) from surface coating of sputtered material from the puller electrode. This avoids arcing and creep currents across the insulation ceramics and allows long-term ion extraction.

The sputtered material is deposited on surfaces exposed directly or indirectly in line of sight to the surface undergoing sputtering. Experience with the first extraction setup from the University of Giessen showed that deposition in line of sight from the sputtered surface dominates rather than indirect deposition around corners. Concentrating our efforts on the reduction of direct surface coating allows to shield the exposed surfaces on the insulation ceramics from line of sight access of the sputtering center. A baffle has been implemented to facilitate this shielding.

This method proved to be very well suited to such an extent that neither difficulties with excessive leakage current nor any visible surface coating of the insulation ceramics have been observed with the new design. Figure 4.17 shows a section cut where the baffle is installed.

4.3.3 Einzel lenses

After extraction the ions drift toward the first and second Einzel-lens (see Figure 4.18). The Einzel-lens focuses the ion beam and shape its emittance to match the acceptance of the consecutive transmission such as the ninety degree mass separation magnet. The first Einzel-lens incorporates the Puller electrode, thereby guaranteeing a fixed distance between the point where the ions are finished accelerating (after extraction) and the point of Einzel-lens entry. This significantly eases experimental manipulation and optimization of SWISSCASE. To match the focus length of the first Einzel-lens with the extraction the hole assembly can be slid forward and backward (see Figure 4.18). In addition the second Einzel-lens can be positioned independently from the first Einzel-lens.

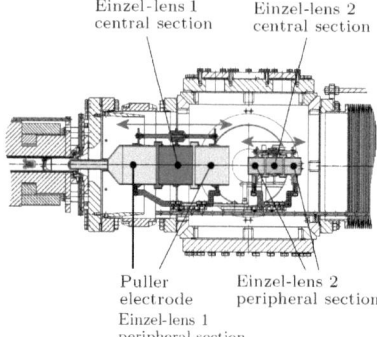

Figure 4.18: Composition of Einzel lenses in SWISSCASE. The puller electrode is attached at a fixed distance to the first Einzel lense. Both Einzel lenses can be moved independently in axial direction (red arrows) by external controls.

4.3.4 Ion beam transmission

The ion beam transmission is responsible for the efficient guidance of the ion beam from the point of extraction through different vacuum vessels, a ninety degree mass separation magnet and finally to the Faraday cup (see Section 7.2.2). The main goal in the design and operation of the transmission system is to minimize ion bean loss. Such loss mainly consists of part of the ion beam colliding with wall elements of the transmission system or with the elements of the ion optics.

The ECR ion source magnet has been tested for plasma production and ion extraction at the University of Giessen by Bodendorfer et al. During these tests an older extraction setup and mass separation magnet from a 14 GHz ECR ion source was used to test the ion beam performance of the future SWISSCASE confinement magnet. To detail the temporal evolution of the different setups a chronological summary is given:

- July 2005 to August 2005: Testing of magnetic confinement and ion beam performance on experimental ECR extraction and mass separation system at the University of Giessen.

- August 2005: Integration of magnetic confinement and extraction system from the University Giessen into test bench in Bern.

- **November, 28th, 2006:** First beam without mass separation, 480 μA, wit CO_2 gas operation and 580 μA with Helium gas operation.

- **February, 12th, 2007:** First beam with mass separation magnet.

- January 2007 to August 2007: Experimentation with plasma parameters, ion optical setup. Acquisition of first spectra.

- August 2007 to February 2008: Bremsstrahlung measurement.

- January 2008: Integration of new vacuum facility, new microwave coupler and new extraction system.

- **January 24th, 2008:** First beam with new setup.

- **January to April 2008:** Acquisition of ion beam spectra.

Starting up the newly installed ECR ion source SWISSCASE in Bern under new conditions such as temperature, microwave and vacuum setup, we tried to avoid unnecessary changes to the already proven extraction mechanism which was retained from the University of Giessen resulting in 'First Light' at the University of Bern at November, 28th, 2006. This setup was not supposed to be final as it did not provide the necessary vacuum quality nor the desired experimental reproducability. But it allowed to confirm the successful operation of the ECR ion source in a new environment with a new microwave generator, new wave guide system and a new vacuum pump setup. With this setup 580 μA of Helium ions were extracted from the ion source measured at the Faraday cup without a mass separation magnet in place.

Spectral measurements in June 2007, with the mass separation magnet in place, resulted in a charge state distribution presented in Figure 4.19. In this spectrum, peaks for He^+, He^{2+} and H^+ are visible. The sum of all these peak currents amounts to 552 μA. This current is 95.2 % of the total current measured without the mass separation magnet in place.

In addition to the comparison of current measurements presented above another indication points toward a highly optimized ion transfer. If the extracted ion beam hits a solid surface two different effects can be observed. Coating patterns are visible where the ion beam hits a solid surface if the energies of the particles is suitably small and the resulting neutralized atoms or molecules form molecular layers and a deposition process is established. If the particle energies are larger, after extraction, or if the resulting neutrals form gases that do not deposit, sputtering patterns are observed (see puller electrode Figures 4.16a-d). Apart from the extraction zone, none of the exposed surfaces of the ion optics nor of the mass separation magnet showed any sign indicating sputtering or deposition. This again shows that the ion beam transmission is highly efficient. At this point no further investigation is performed in the improvement of the ion beam transmission of SWISSCASE.

We can conclude from this Section that:

- the transmission of the ion beam for Helium ions is highly efficient (95.2%) compared to the not mass discriminated total extracted current from a previous measurement. No surface coating patterns nor sputtering is observed on exposed parts. This suggests that the ion optics is close to its optimal design. The design parameters of the Einzel-lens and extraction geometry do not need to be modified with respect to the design obtained from the University of Giessen.

4.4 Radiation shielding

Radiation from the ECR plasma can be harmful to human beings, and the system must therefore be shielded to protect the operator. The Bremsstrahlung originates from plasma electrons colliding with plasma ions, neutrals and the plasma chamber wall. The photon spectrum containing energies of the X-ray

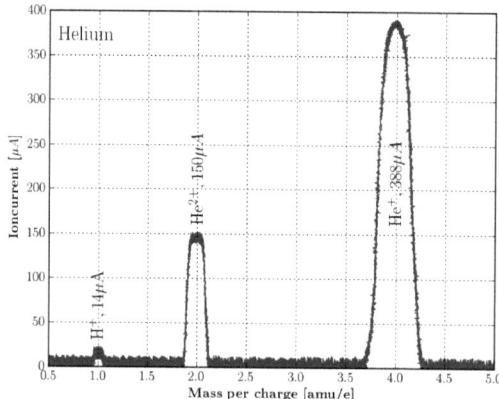

Figure 4.19: Helium spectrum acquired with original extraction setup. Parameters are optimized for He^{2+}: $f = 10.50$ GHz, $P = 50$ W, $p_3 = 8.0 \cdot 10^{-5}$ mbar, $U_{ex} = 4$ kV. The sum of all currents shown in this spectrum is 552 μA.

spectrum. X-ray photons have a rather low effective cross section with organic molecules found in the human body. However if an interaction occurs the X-ray photon is able to ionize and disintegrate organic molecules of all sizes including proteins and nucleic acids of human DNA. This effect can lead to a change in body chemistry, destruction of cells hit by the photon or in rare cases to stimulation of the cell to degenerate and become a cancer cell. For this effect there is no minimum radiation dose which can be considered to be harmless. However for reference there is a natural back ground radiation level to which the human body is exposed at all times depending on location and daytime. The average back ground radiation dose rate level in Switzerland is 0.1 μS per hour, corresponding to 0.876 mS per year.

All measures taken to reduce the radiation dose load taken from the presence of the ECR ion source are based on the background radiation dose as a radiation reference level. With no measures in place SWISSCASE induces radiation dose rates up to 10 μS/h at a working distance of 0.2 m. This is 100 times the background radiation dose rate. Hence measures were taken to reduce the measured radiation level to the background value as a target level.

X-rays and Gamma rays are absorbed in a given shielding material according to Eq. (4.1).

$$\frac{I}{I_0} = exp(-(\mu/\rho)\Delta d) \qquad (4.1)$$

I_0 is the incident intensity of the radiation without shielding material in place, I is radiation intensity at the same distance with shielding material of thickness Δd, of mass density ρ and attenuation coefficient μ in place.

Our own bremsstrahlung measurements resulted in a photon energy distribution peaking at 50 keV. μ/ρ for lead at this energy is 8.041 cm^2/g [30]. Lacking the knowledge about the fraction of high energy photons up to 1 MeV we decided to implement an amount of lead which is practical to handle during assembly and disassembly and which represents no critical mechanical load to the structure. These considerations resulted in a minimal thickness of 20 mm. At a photon energy of 50 keV this leads to an intensity ratio of $I/I_0 = 6 \cdot 10^{-80}$. However at a photon energy of 1 MeV the corresponding absorption coefficient of lead at this energy is 0.07102 cm^2/g [30] leading to an intensity ratio of $I/I_0 = 0.1997$. The dominating part of the measured intensity is contributed by photons with an energy around 50 keV (see Chapter 8). X-ray Measurements with the shielding installed and ECR plasma active, resulted in no detectable radiation above background. Hence the shielding is considered to be safe also for long term working exposure.

4.5 Vacuum setup and gas feed

SWISSCASE is operated in a pressure range from $8 \cdot 10^{-4}$ $mbar$ to 10^{-9} $mbar$. To determine the flow regime we can calculate the Knudsen number for this pressure range. The temperature is supposed to be ambient temperature 20 C°. The Knudsen number is then given by Eq. 4.2.

$$K_n = \frac{\lambda}{L} \quad (4.2)$$

K_n is the Knudsen number, λ is the mean free path of the particles of interest and L is a characteristic length of the system, the flow tube diameter for instance. We have the following criteria [40, 2] (Table 4.5).

$K_n > 1/2$	Free molecular flow
$1/2 \geq K_n \geq 0.1$	Transition flow ($\lambda \cong L$)
$0.1 > K_n$	Continuum flow

Table 4.5: Summary of criteria to distinguish free molecular flow, transition flow and continuum flow.

The mean free path λ can be calculated according to Eq. 4.3 [7, 11].

$$\lambda = \frac{1}{n\sigma} \quad (4.3)$$

n is the particle number density and σ the inter particle cross section. During operation n is $\cong 10^{19}$ $1/m^3$ [4] and $\sigma \cong 10^{-20}$ m^2. This leads to $\lambda \cong 10$ m. For L, the characteristic length, we can take a typical diameter such as the inner diameter of the plasma chamber (24 mm). This leads to $K_n \cong 417$. Even if we take for L the overall length of the facility ($L = 1499$ mm), we get $K_n \cong 6.7$ which is still in the free molecular flow region. Hence inter particle collisions can be neglected and the flow behavior is dominated by free molecular flow behavior.

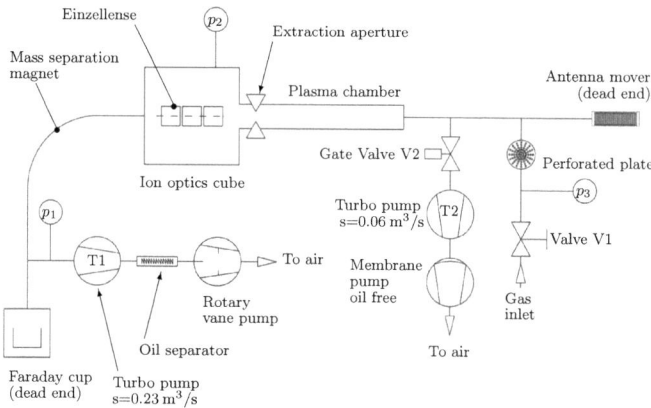

Figure 4.20: Scheme of vacuum and gas flow setup. p_1, p_2 and p_3 denote pressure sensors. The oil separator protects the UHV part of the facility from oil back streaming of the rotary vane pump during an emergency shutdown of the turbo pump T_1. This precaution is not necessary for turbo pump T_2 because the membrane pump is oil free.

Figure 4.20 gives an overview of the gas feed and vacuum system. The operation gas enters the UHV facility through the thermal electrical valve V1. The gas passes through a perforated plate, which serves as a microwave reflector and absorber. The gas then enters a volume connecting the dead end of the microwave antenna mover, the T-connector to the small turbo pump and the plasma chamber. A certain fraction of the gas flow will go into the small turbo pump and will leave the system. The remaining flow fraction will enter the plasma chamber and serve as working gas for the ECR plasma operation. After plasma reaction the remaining gas flows through the aperture entering the ion optics cube. From there the gas flows through the 90° mass separation magnet to reach the large turbo pump to be evacuated from the system.

4.5.1 Vacuum capabilities

With the inlet gas valve closed the facility routinely demonstrates a vacuum pressure of $5 \cdot 10^{-9}$ mbar measured by the pressure sensor p_2 at the ion optics cube. Under the same conditions the vacuum pressure measured by the pressure sensor p_3 near the gas inlet is as low as 10^{-8} mbar. However after heating the vacuum facility to 60 C° for one week the pressure measured by the pressure sensor p_2 at the ion optics cube dropped below the measurement range limited to 10^{-9} mbar entering the range of 10^{-10} mbar. With the presently installed pressure sensors (Balzers, Compact Full Range Gauge, PKR 260) we are unable to resolve this vacuum pressure regime.

During operation the ion spectra show traces of air molecules. The air

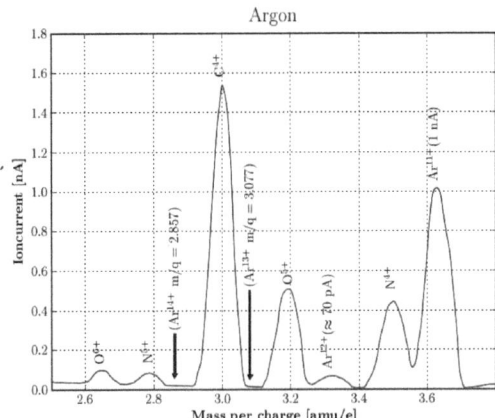

Figure 4.21: Ion spectrum of SWISSCASE. Operation of an Argon plasma (100 W, 10.88 GHz). The operation parameters are optimized for an ion current of Ar^{11+} and Ar^{12+}. Despite the contamination with Oxygen, Nitrogen and Carbon the detection of possible occurrence of Ar^{13+} and Ar^{14+} is not impeded because the peaks of interest do not cover the expected locations of Ar^{13+} and Ar^{14+}.

contamination is not limiting the resolution of high charge state ions such as Ar^{8+}. Figure 4.21 shows a spectrum produced from an argon plasma optimized for the highest charge states possible in this facility Ar^{11+} and Ar^{12+}. This operation regime is the most vulnerable to contaminations of any kind and is therefore a good benchmark for the vacuum quality because highly charged ions undergo charge-exchange reactions with the background gas and thus are converted to lower charged ions. We see the presence of (Ar^{11+} and Ar^{12+}) and a contamination with O^{6+}, N^{5+}, C^{4+}, O^{5+} and N^{4+}. The presence of carbon in the shown spectrum originates from a former CO_2 plasma operation.

We can conclude from this Section that:

- the vacuum facility is able to routinely provide a clean vacuum with residual pressure of $5 \cdot 10^{-9}$ mbar without the need of additional heating. By heating the facility to 60 C° for one week the residual pressure can be improved to $1 \cdot 10^{-9}$ mbar. The contamination of air components and carbon is present in an expected amount but not impeding the operation of SWISSCASE nor of the mass separation process.

4.5.2 Plasma chamber gas pressure

To determine the gas pressure in the plasma chamber we can calculate the different gas flow resistances given by the tubing and the implemented orifices, then backtrack the pressure profile from the gas inlet measured by the pressure

58 Chapter: Realization of source elements

sensor p_1. The tubing connecting the gas inlet valve V_1 with the plasma chamber represents a rather sophisticated geometry for a free molecular flow because of the presence of the microwave antenna, the flow bifurcating turbo pump T_1, the right angle connector and the microwave coupling cube. Applying formulas from literature will lead to a rather inaccurate result requiring numerical simulation. However we can also access the plasma chamber pressure in an another way avoiding the mentioned difficulties.

As for most turbo pumps operated in this pressure regime, also the volume flow rate of the turbo pump T_1 can be considered pressure independent [40]. We know the volume flow rate of turbo pump T_1 to be $q_V = 0.23\ m^3/s$ from specifications and we can measure the pressure at this turbo pump inlet by p_1. This allows to calculate the pressure volume flow rate (Eq. 4.4).

$$q_{pV} = q_V \cdot p_1 \qquad (4.4)$$

In addition q_{pV} can be related to q_n, the particle flow rate, by the ideal gas law (Eq. 4.5)

$$q_n = \frac{q_{pV}}{Tk_B} \qquad (4.5)$$

T is the temperature of the gas and k_B the Boltzmann constant. As leakage measurements showed, the leakage flow rate of the facility is well below $2.3 \cdot 10^{-10}$ mbar $\cdot$ m^3/s at the pressure level we operate the ECR plasma. Hence we can take q_n as a constant value along a stream line from p_1 to the plasma chamber. Further assuming the temperature of the gas does not change along the same streamline allows q_{pV} to be constant along the stream line.

We can measure the pressure at p_2 during plasma operation. We further know exactly the geometry of the extraction aperture acting as a flow restricting element. The plasma chamber pressure p_{plasma} is therefore given by (Eq. 4.6).

$$p_{chamber} = p_1 + \Delta p_{aperture} \qquad (4.6)$$

$\Delta p_{aperture}$ is the pressure drop caused by the gas flow streaming through the extraction aperture. This pressure drop can be calculated by (Eq. 4.7) [40]

$$\Delta p_{aperture} = \frac{4 q_{pV}}{\bar{c} \cdot A \cdot p_{aperture}} \qquad (4.7)$$

$\bar{c}$ is the mean velocity of the gas particles, A is the cross section of the extraction aperture and $p_{aperture}$ is a transmission probability depending on the geometry only. $p_{aperture}$ is given by (Eq. 4.8)

$$p_{aperture} = 1 - \frac{1}{2}\frac{l}{r} = 0.483 \qquad (4.8)$$

l is the length of the aperture orifice and r its radius. Further we calculate $\bar{c}$ by (Eq. 4.9)

$$\bar{c} = \sqrt{\frac{8RT}{\pi M_{Molar}}} \qquad (4.9)$$

R is the universal gas constant, T the temperature and M_{Molar} the molar mass of the gas. Table 4.6 summarizes the data for a typical Ar plasma operation.

Entity	Origin	Value
p_1	Measured	$1.1 \cdot 10^{-6}$ $mbar$
q_V	Specifications Balzers	0.23 m^3/s
q_{pV}	$q_V \cdot p_1$	$2.5 \cdot 10^{-7}$ $mbar \cdot m^3/s$
p_2	Measured	$1.3 \cdot 10^{-5}$ $mbar$
T	Measured	300 K
R	Weast at. al [37]	8.314472 $J/K \cdot mol$
$M_{Molar-Ar}$	Weast at. al [37]	$39.95 \cdot 10^{-3}$ kg/mol
$\bar{c}$	Wutz et al. [40]	398.633 m/s
$P_{aperture}$	Eq. 4.8	0.4383
A	$A = r^2 \pi$	$2.827 \cdot 10^{-5}$ m^2
$\Delta p_{aperture}$	Eq. 4.7	$1.9 \cdot 10^{-4}$ $mbar$
$p_{chamber}$	$p_2 + \Delta p_{aperture}$	$2 \cdot 10^{-4}$ $mbar$

Table 4.6: Summary of measured and calculated values.

4.5.3 Discussion

To verify the obtained value of the plasma chamber gas pressure we can compare the corresponding gas density at room temperature and the ECR plasma density given by the criteria that the plasma is opaque for the incident microwave frequency (see Chapter 3). This is an approximation because the gas density depends on the gas pressure and the gas temperature. The latter is above room temperature because of the interaction with the hot plasma. However the long mean free paths in this pressure regime do not lead to many collisions between neutral gas atoms and plasma ions or electrons as shown in Chapter 3. The overall momentum transfer from the plasma to the neutral gas is therefore low and the background gas can be considered to be at room temperature for a first approximation.

We calculate the room temperature density $n_{neutral}$ of the plasma chamber gas pressure by (Eq. 4.10)

$$n_{neutral} = \frac{p_{chamber}}{k_B T} = 4.8 \cdot 10^{18} \ 1/m^3 \qquad (4.10)$$

The plasma density limited by its transparency with respect to the incident microwave is given by (Eq. 4.11)

$$n_{plasma} = \frac{f^2 \epsilon_0 m_e}{e^2} = 1.461 \cdot 10^{18} \ 1/m^3 \qquad (4.11)$$

f is the microwave frequency (10.88 GHz), ϵ_0 is the permetivity of free space, m_e is the electron mass, and e the electron charge. We can now give an approximation of the ionization fraction by (Eq. 4.12):

$$\eta = \frac{n_{plasma}}{n_{neutral}} = 0.3 \qquad (4.12)$$

This result is in good agreement with literature. The ionization fraction of 2.54 GHz ion sources gathers clusters 10% [11, 17, 26]. SWISSCASE operates at a frequency of 10.88 GHz and is able to produce a larger degree of ionization than 2.54 GHz ion sources [17, 26]. Higher charged ions require a reduced chance of recombination with neutrals and hence a higher fraction of ionization. Hence we expect a higher ionization fraction from SWISSCASE compared to 2.54 GHz ion sources.

4.6 The high voltage setup

SWISSCASE requires high voltage supply for the extraction setup and two independent Einzel-lenses. Figure 4.22 gives an overview of all parts connected to high voltage

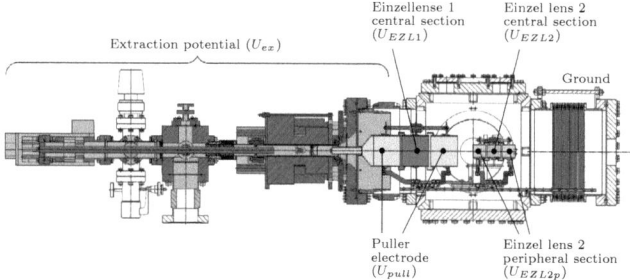

Figure 4.22: High voltage parts in section cut showing a side view of SWISSCASE. Insulation is shaded orange. Not colored parts are grounded.

During ion beam operation all high voltage potentials can be adjusted independently. However practical experience showed the following potential settings to be of interest (Table 4.7).

Potential	[kV]
U_{ex}	0 to 12
U_{pull}	0 to -2.5
U_{EZL1}	0 to 10
U_{EZL2}	0 to 2
U_{EZL2p}	ground

Table 4.7: Potential ranges of interest.

In the following Sections we discuss in more detail the potential distribution and its implications with the plasma and the ion beam formation.

4.6.1 High voltage for extraction

To extract the ions from the ECR plasma high voltage is applied to the extraction setup (for mechanical details see Section 4.3.1). The applied extraction voltage leads to a potential difference creating a potential field extending into the plasma as shown in Figure 4.23.

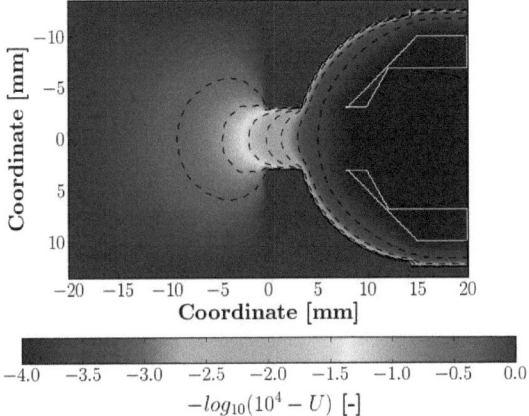

Figure 4.23: Simulation of the potential distribution in the extraction region. U is the potential in Volts. For better visualization the function $-log_{10}(10^4 - U)$ is displayed rather than U. The shown equipotential lines (dashed) are equally spaced in log-space. This transforms to lines at: $9999\,V$, $9997\,V$, $9990\,V$, $9968\,V$, $9900\,V$, $9683\,V$, $9000\,V$ and $6837\,V$. The potential field extends into the plasma chamber to extract plasma ions trapped by the magnetic field and the counteracting plasma potential. The simulation does not take into account the presence of the plasma conductivity which leads to a reduction of potential differences inside the plasma chamber (to the left in the diagram).

Figure 4.23 shows an electric field calculation of the potential drop inside the plasma region that is small compared to the acceleration region to the right side of the illustration. However according to Melin et al. [14] the plasma is charged slightly negative ($\cong -3\,V$). Hence ions are trapped not only by the magnetic confinement field but also by the negative plasma potential created by the ECR electrons. This negative potential needs to be overcome for the ions to escape the plasma. This shows that also small field residuals from the extraction setup extending into the plasma are intense enough to pull out plasma ions.

The plasma potential superimposes the extraction potential. This defines a surface where the isocontour surfaces of the extraction potential matches the plasma potential. This virtual surface gives a first approximate location of the ion well, called the extraction meniscus. The extraction is further modified both in shape and location by the plasma pressure, the magnetic field and space charge created by the extracted ions discussed in more detail in Section 6.

Due to the applied potential difference necessary for the ion extraction, locally high electric fields arise at specific locations. Critical electric field densities occur at the tip of the puller electrode and at its shoulder. Figure 4.24 gives an overview of the electric field situation.

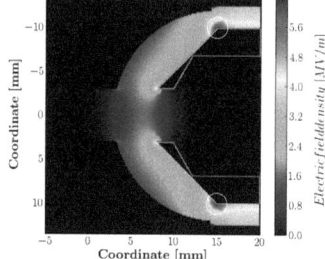

Figure 4.24: Simulation of the electric field at the puller electrode. The highest electric fields are present near the shoulders rather than at the tip of the puller electrode (see white circles). Despite the high electric field density at the shoulder, arcing does not occur unless the puller electrode is extended maximally toward the extraction aperture.

Despite the presence of high electric fields no visible erosion on the puller electrode has been observed after plasma operation in various operation regimes and with different extraction potentials. However arcing occurs during plasma operation if the puller electrode is extended maximally toward the extraction aperture as shown in the field calculation of Figure 4.24. In this position the puller electrode does not yield a favorable ion current. Hence arcing due to the position of the puller electrode is not considered to be problematic as it does not occur at any other position of the puller electrode.

A significant part of SWISSCASE has to be insulated from ground to apply the extraction voltage. Figure 4.25 shows the parts connected to high voltage and the parts which are grounded.

During manipulation or disassembly of the extraction unit great care has to be taken to ensure that no contamination with finger prints or graphite pens are made on the ceramic insulator. Any such contamination could lead to an excessive leakage current and the consecutive emergency shutdown of the high voltage system either by the current limiter or the vacuum protection system.

4.6.2 High voltage setup of ion optics

The ion optics consists of two separate, isolated Einzel-lenses. Each Einzel-lens can be connected to a different high voltage potential. Einzel-lens 1 includes the puller electrode responsible for the ion extraction described in the previous Section. This design brings the advantage of defining the central Einzel-lens section potential relative to the puller electrode potential rather than to ground potential if needed. It further guarantees a constant geometry between the puller electrode tip and the Einzel-lens 1 for different puller electrode positions

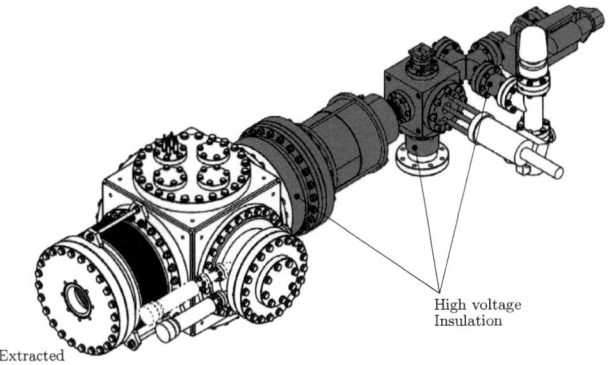

Extracted ions

High voltage Insulation

Figure 4.25: 3D view of high voltage parts in red and insulation parts in magenta. White parts are grounded.

relative to the extraction aperture. This significantly simplifies ion optical adjustments during testing and operation with different puller electrode positions and potentials. Figure 4.26 shows a simulation of a typical potential distribution along the z-axis of SWISSCASE during plasma operation and ion beam extraction.

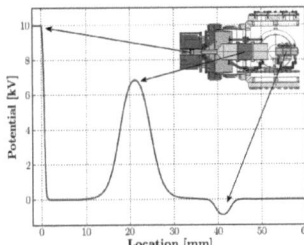

Figure 4.26: Calculation of typical potential distribution along z-axis. Arrows are relating the locations of interest to a section view. In this simulation the extraction potential is set to 10 kV. The puller electrode and the peripheral section of Einzel-lens 2 are grounded. Einzel-lens 1 is set to 7.5 kV and Einzel-lens 2 is set to -1 kV.

The high voltage design parameters are based on the optimal beam energy of 10 kV with respect to the ion current performance (see Broetz and Trassl et al. [6, 36]). In addition, simulations with TOSCA and SCALA showed the corresponding optimal Einzel-lens voltage to be 7.2 kV to 10.5 kV depending on the extraction potential difference $U_{ex} - U_{pull}$, puller electrode position and the extracted ions. Table 4.8 summarizes the voltage conditions at which the construction was successfully tested. Higher voltages especially for U_{EZL1} seem feasible due to the specific vacuum gaps but still need to be tested.

Potential	[kV]
U_{ex}	-10 to 10
U_{pull}	-10 to 10
U_{EZL1}	-15 to 15
$U_{ex} - U_{pull}$	$\leq 15\ kV$
U_{EZL2}	-10 to 10
U_{EZL2p}	-10 to 10
$U_{EZL2} - U_{EZL2p}$	$\leq 2\ kV$

Table 4.8: Tested voltage limits.

4.7 Mass separation magnet

As part of the SWISSCASE ECR ion source project a 90 degrees bending magnet is required to separate particles of different mass/charge ratios. The characterizing attribute of such a mass separation magnet is the path integral of the magnetic field along the path length along which the particle will travel:

$$\alpha = \int d\alpha = \int \frac{dl}{r} = \int \frac{qB(l)}{vm} dl = \frac{q}{p} \int B(l) dl \qquad (4.13)$$

α is the total deviation angle of the particle with charge q, mass m, velocity v, momentum p, traveling through a magnetic field B for a path length of l. The specific calculation is done by the concept to keep the heaviest ion with the least possible charge on a circular path for an angle of 90 degrees. In a first brief calculation the classic and the relativistic momentum will be compared to justify the usage of the classic momentum. This special calculation is done with the mass of a proton, rather than the specified ^{131}Xenon as this will be the fasted particle we will measure and keep track on inside the magnet. The symbols used in this calculation are in MKS units as follows (Table 4.9):

4.7.1 Non relativistic justification

First a brief calculation shows the ratio between the classic and the relativistic momentum leading to the conclusion that using the classic momentum is justified. The most lightweight and hence fastest particle which we will accelerate will be a proton. The classic momentum of the accelerated proton is Eq. (4.14):

$$p_{classic} = m_0 v = \sqrt{2E_{kin}m_0} = \sqrt{2Uqm_0} \qquad (4.14)$$

The relativistic momentum of the accelerated proton is Eq. (4.15):

$$p_{relativistic} = \frac{1}{c}\sqrt{E_{kin}(E_{kin} + 2E_0)} \qquad (4.15)$$

We can write Eq. (4.16):

$$\frac{p_{classic}}{p_{relativistic}} = c\sqrt{\frac{2E_{kin}m_0}{E_{kin}(E_{kin} + 2E_0)}} = c\sqrt{\frac{1}{\frac{E_{kin}}{2E_0} + 1}} = \frac{1}{\sqrt{\frac{Uq}{2m_0c^2} + 1}} \qquad (4.16)$$

Entity	Description	Value	Unit
B	Magnetic field density		$[T]$
α	Bending angle	$\pi/2$	$[rad]$
L	Path length in homogenous B-field		$[m]$
U	Acceleration voltage	10^4	$[V]$
m_{Xe}	Rest mass of heaviest particle (Xe_{131})	$2.17 \cdot 10^{-25}$	$[kg]$
m_0	Rest mass of proton	$1.66 \cdot 10^{-27}$	$[kg]$
q	Particle charge		$[C]$
F	Force required to keep particle on track		$[N]$
a	Acceleration of particle		$[m/s^2]$
r	Radius of circular path		$[Ns]$
t	Time in B-field		$[s]$
p	Momentum of particle (Xe_{131})		$[Ns]$
E_{kin}	Kinetic energy of particle		$[J]$
E_0	Rest energy of particle	$E_0 = m_0 c^2$	$[J]$
c	Speed of light in vacuum	$2.997924458 \cdot 10^8$	$[m/s]$

Table 4.9: Entities, values and units used for the calculation of the path integral necessary for the determination of the mass separation magnet.

For given values we obtain: $\frac{p_{classic}}{p_{relativistic}} = 1 - 2.68 \cdot 10^{-6}$. The deviation from $p_{classic}$ is therefore small enough to carry on with the classic momentum.

4.7.2 Calculation of magnetic path integral

In the following we calculate the necessary magnetic path integral (Eq. 4.13) to keep a charged particle on track for a bending angle of $\pi/2$. Figure 4.27 shows a scheme for the performed calculation. Please note for this approximation fringing fields at the ends of the magnetic field, the entry and the exit point of the particle, have been neglected. This approximation allows to calculate a sufficiently precise value for the necessary magnetic path integral.

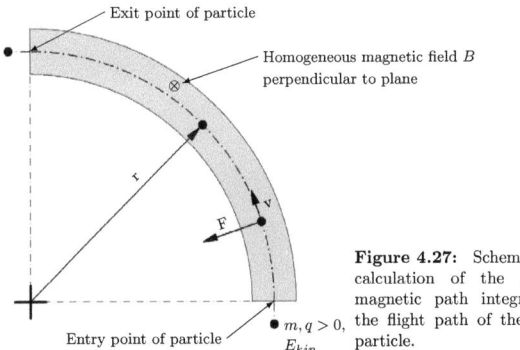

Figure 4.27: Scheme for the calculation of the necessary magnetic path integral along the flight path of the charged particle.

For simplification we assume the magnetic field density along the path length to be homogenous. Then the force required to keep the charged particle on track in the homogeneous magnetic field is given by Eq. (4.17):

$$F = a \cdot m_{Xe} = \frac{v^2}{r} \cdot m_{Xe} = B \cdot q \cdot v \qquad (4.17)$$

The path geometry allows to write Eq. (4.18):

$$a \cdot r = L \qquad (4.18)$$

We can combine Eq. (4.17) and Eq. (4.18) to Eq .(4.19):

$$\frac{a \cdot v^2 \cdot m_{Xe}}{L} = B \cdot q \cdot v \qquad (4.19)$$

Simplifying leads to Eq. (4.20):

$$L \cdot B = \frac{a \cdot p}{q} \qquad (4.20)$$

Which is the value of the path integral we are looking for. With the given data we obtain a value of: $L \cdot B = 0.25882$ Tm. This value allows to chose a suitable bending magnet which satisfies the required path integral.

4.7.3 Choice and purchase of mass separation magnet

Different inquiries resulted either in cost, volume, mass or delivery times unacceptable for this project. The University of Giessen however offered one of their used mass separation magnets including current driver which has been used to analyze beams of Xe_{131} at an extraction potential of 10 kV. In our own experiments the mass separation system operates at a maximum of 0.67 T to analyze a beam of Xe_{131} extracted at 10 kV. The current supply is the limiting device preventing any further increase in magnetic field density. However this magnetic flux density is not sufficient to saturate the iron material of the mass separation magnet. Hence also particles with an even larger fraction of kinetic energy per charge could be resolved for spectral analysis given a more powerful current supply.

Chapter 5

Numerical Simulation of MEFISTO

In this chapter the numerical simulation of the MEFISTO ECR ion source located at the University of Bern will be presented.

5.1 Introduction

The magnetic confinement of both MEFISTO and SWISSCASE are composed of permanent magnets only. The whole ECR ion source is elevated to high voltages up to 100 kV for post acceleration of the extracted ions. No solenoid coils are used in favor of a low electric power consumption for the operation on the high voltage terminal. In addition the use of permanent magnets instead of solenoids brings different advantages such as compact size and no cooling requirements. However this comes at the cost of giving away the option of magnetic field adaptation after installation.

Unlike the magnetic field of solenoid coils the field of permanent magnets can only be tuned over a very limited range after the manufacturing process and the construction of the ECR ion source. Great care has therefore to be taken in the design of the magnetic configuration because adaptation of the magnetic field after construction is limited to very few options. The magnetic arrangement chosen for the MEFISTO [4] ion source features a field distribution which cannot be reduced from three to two dimensions for the purpose of a more economic simulation. Hence full 3D simulations have to be performed to visualize and understand the full spatial extent of the field distribution.

This chapter presents the MEFISTO finite element model simulation results of the magnetic field distribution and the numerical electron trajectory integration. The result of the magnetic field simulation is compared to measured data validating its accuracy. The simulation of the electron trajectories seems plausible but cannot be compared to direct measurements because the experimental determination of electron trajectories is not possible yet in SWISSCASE. However observed surface coating patterns match the simulated electron densities, supporting the simulation accuracy.

In the next Section a brief overview is given of the principle of calculation of the finite element solver used for all finite element model simulations presented in this thesis. Section 5.3 presents the finite element model used to simulate the magnetic field. Section 5.4 shows the results of the FEM simulation followed by Section 5.5 revealing the results of the numerical electron trajectory integration. Section 5.6 concludes the finding in this chapter.

5.2 Finite element solver

TOSCA by Vectorfields is a commercial high performance finite element solver with the ability to simulate magnetic and electric fields. TOSCA performs numerical iterations until a predefined convergence criteria is fulfilled. With this method TOSCA can simulate permanent magnetic setups made of material with nonlinear properties specified by the user. The name of TOSCA symbolizes its principle of operation and stands for total scalar potential.

Instead of solving for the three dimensional vector magnetic field B, TOSCA uses a scalar potential to solve the set of finite element equations. To clarify this we have a look at the fourth Maxwellian equation in vacuum (5.1).

$$\nabla \times \vec{B} = \mu_0 \vec{j} + \mu_0 \epsilon_0 \frac{\partial \vec{E}}{\partial t} \qquad (5.1)$$

$\vec{B}$ is the magnetic field, $\vec{E}$ the electric field, ϵ_0 the permittivity of vacuum, t the time, μ the permeability of vacuum and $\vec{j}$ the current density.

In general the right hand side of Eq. 5.1 is not zero. However as a good approximation for our special case of a permanent magnetic setup there are no current densities $\vec{j}$ and no electric field $\vec{E}$ changing with respect to time to be considered. Hence we can nullify the right side of Eq. 5.1 and we can state the following reasoning:

$$\nabla \times \vec{B} = 0 \quad \Rightarrow \quad \exists \phi : \nabla \phi = -\vec{B}$$

B represents a conservative field due to its nullified curl. Hence there is a potential ϕ for $\vec{B}$. This allows to use a scalar to solve the finite element problem instead of a three dimensional vector which makes numerical simulations much more efficient.

In the next section the application of this concept on the finite element model of the MEFISTO ion source located at the University of Bern is demonstrated.

5.3 Magnetic finite element model of MEFISTO

Figure 5.1 shows the arrangement of the permanent magnets in the MEFISTO ion source. The six permanent magnets of the ring-shaped cluster on the left are all magnetized inwards. Similarly the cluster on the right side is magnetized outwards. These two outer ring-shaped clusters form the magnetic field of an axial magnetic bottle and thereby establish the axial confinement of the plasma electrons. In contrast the central magnet cluster consists of permanent magnets which are magnetized alternatively inwards and outwards. This central cluster forms the hexapole field which produces the radial confinement of the plasma electrons. The ECR plasma is located inside this central ring arrangement. In

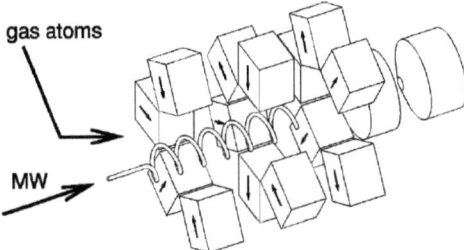

Figure 5.1: The arrangement of the full permanent magnets of the MEFISTO ECR ion source at the University of Bern [Marti et al, 2001, [26]]. Some magnets are removed in this drawing for better visualization.

addition Figure 5.1 also shows the helix antenna which radiates the microwave power into the ECR plasma.

As discussed in Chapter 3 the choice of the excitation frequency band determines the required magnetic field density B. Hence the design parameters of the magnet structure have to be chosen carefully so that the magnetic field fulfills the ECR criteria (3.1). This criteria is generally fulfilled on a closed isosurface of the magnetic field density (see Section 5.4). This isosurface defined by the magnetic field distribution and the microwave frequency has to be located entirely inside the plasma container for optimal ECR efficiency.

This ECR ion source is used for the calibration of spacecraft instruments which are designed to detect highly charged ions in space. The performance of the ECR ion source with respect to the production of these highly charged ions depends on the square of the ECR resonance frequency [11]. In addition the magnetic field distribution forms a magnetic mirror with a ratio $r_m = B/B_{\text{centre}}$ in all directions, which should be as high as possible for optimal electron and ion confinement. However the successful extraction of highly charged ions requires a locally poor confinement performance so that the ions can be extracted from the loss cone in the axial beam direction. In Figure 5.1 this extraction system is formed by the two cylinder shapes to the right.

All of these requirements have to be met for the successful operation of an ECR ion source. To achieve these goals three dimensional numerical simulations of the magnetic field can be used which allow the highly accurate design of future ECR ion sources. The simulation technology has been tested and the results have been compared with measurements of the existing MEFISTO ECR ion source.

Due to the unique field distribution of hexapole confined ECR ion sources the high energy electron population features a trident shaped spatial distribution as predicted by numerous authors ([7], [11]). The magnetic field simulation presented in this chapter confirms both theory and experiment in three dimensions.

Magnetic simulations have been performed using *Tosca* by *Vector Fields* (www.vectorfields.com). The resulting finite element mesh itself has been tested extensively with reference patterns and circular charged particle trajectories such as electrons and ions. No deviation from the circular orbits were discovered within the system limit of 5000 integration steps for each particle. A summary of the FEM mesh characteristics is given in Table 6.3.

number of linear elements	$2.4 * 10^6$
number of quadratic elements	$1.9 * 10^5$
number of nodes	$7.2 * 10^5$
number of edges	$3 * 10^5$
RMS	1.033
distortion of worst element	$8.31 * 10^{-4}$

Table 5.1: Some characteristics of the finite element model.

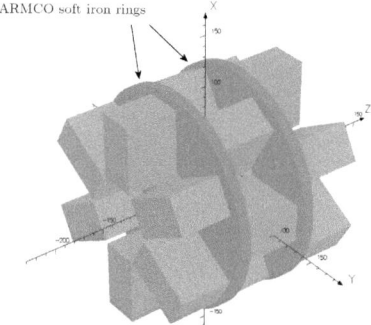

Figure 5.2: The finite element model of MEFISTO with its essential parts. Block magnets in light green and both ARMCO soft iron rings in dark green are visualized.

Figure 5.2 shows the finite element model used to simulate the magnetic field. As described in Section 5.1 the model consists of three sets of block magnets arranged in rings around the central plasma area and two massive soft iron rings separating each of the block magnet ring arrangements. The soft iron rings allow to modify the magnetic field by shortening the magnetic flux. Figure 5.3 shows the magnetic field density along the beam axis with respect to the z-coordinate. The ECR zone is indicated at 87.5 mT where the field density meets the ECR condition.

The two maxima and the miminum of the axial magnetic field of the existing MEFISTO ECR ion source has been measured previously [26] [23]. The local maxima and the local minimum of the measured magnetic field are 240 mT and 60 mT respectively. The corresponding mirror ratio is $r_m = 240$ mT$/60$ mT $= 4$. The maxima of the numerical simulation are 240.1 mT, the minimum is 60.05 mT and the mirror ratio $r_m = B_{max}/B_{min}$ is 3.998. The numerical model is in excellent agreement with the measurement. All following simulations and results are based on the results of the presented numerical field simulation.

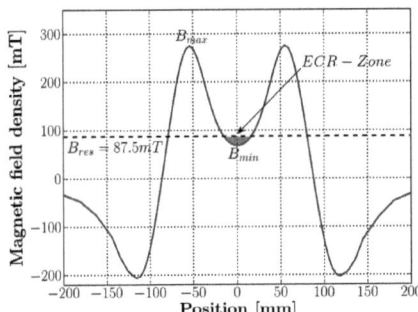

Figure 5.3: Simulated magnetic field along the beam axis.

5.4 The ECR zone

The ECR ion source is fed with a constant microwave frequency of 2.45 GHz, so the location of the resonance region can be determined by the corresponding magnetic field density $B_{res} = 87.5$ mT (3.1). To present a two-dimensional analysis of the field distribution and the ECR locality inside the magnet system we introduce cut planes (Plane A, B, C) into the three dimensional model as shown in Figure 5.4.

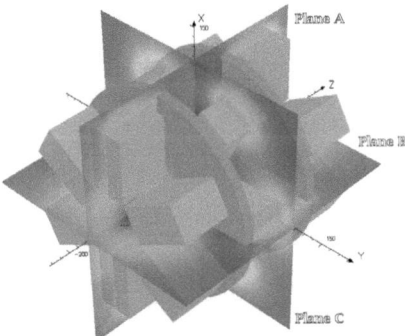

Figure 5.4: To map the magnetic field onto two dimensional patches we introduce cut planes in the 3D model.

In the following figures the magnetic field density has been mapped to color space. There are two figures of each cut plane (see Fig. 6.3 to 6.8) giving an overview and a detailed view of the central part of each plane respectively.

Isocontour lines of the field density are also indicated on each graph.

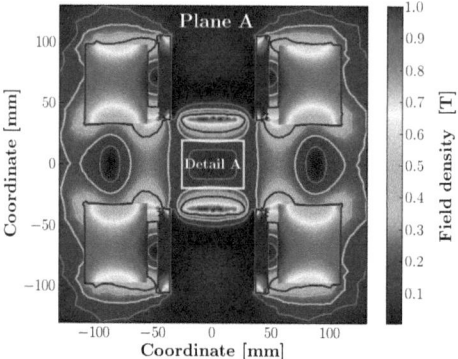

Figure 5.5: Simulated magnetic field in plane A.

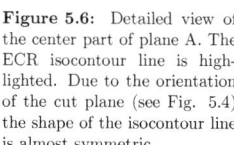

Figure 5.6: Detailed view of the center part of plane A. The ECR isocontour line is highlighted. Due to the orientation of the cut plane (see Fig. 5.4) the shape of the isocontour line is almost symmetric.

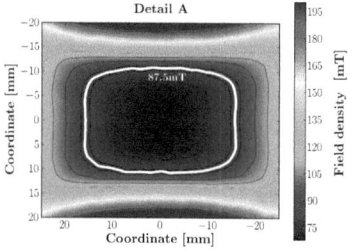

The magnetic field is cut at a different angle to reveal its non symmetric distribution in cut plane B (Figure 6.6). This feature will be visible more clearly in the 3D view presented further down in this Section.

Despite the hexapole arrangement of the central magnet cluster the resulting field distribution in the symmetry plane C (see Fig. 5.4) yields concentric isocontour lines as presented in Figures 6.8. However this is true only for contour lines which are close to the center of the hexapole. Contour lines further from the center and closer to the actual block magnets show the expected hexapole pattern with its distinct poles.

The small irregularities within the contour lines are due to the discretization of the model by the finite element size. The resulting relative error in the field density is lower than 10^{-2}.

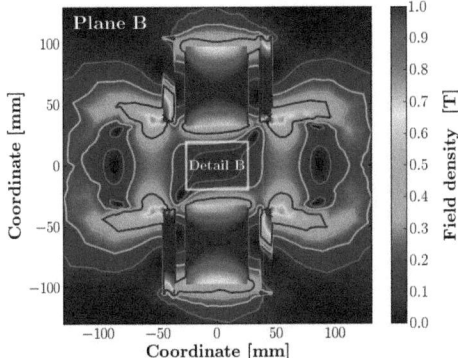

Figure 5.7: Simulated magnetic field density in plane B.

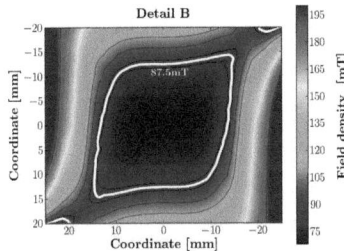

Figure 5.8: Detailed view of the center part of plane B. The ECR isocontour line is highlighted. Cut plane B is rotated by 90 degrees with respect to plane A. The contour lines are no longer symmetric with respect to the beam axis.

In each detailed view the isocontour line of 87.5mT is highlighted. Extending the isocontour line concept from two to three dimensions leads to an isocontour surface. This isocontour surface includes all element centroids of the finite element model which fulfill the criteria of the given magnetic field density $B_{ref} = 87.5$ mT. On this surface the gyration frequency of the electrons in the magnetic field equals the microwave frequency. The plasma electrons are therefore heated effectively on this surface. The surface defined by this criteria is shown in Figure 5.11.

In Figure 5.11, the centroid of each FEM element of the specified surface is represented by a small bubble. Each bubble also represents one launch coordinate for the trajectory model represented in the following section. The shape is point symmetric with respect to the origin located at the center of the shape. Note the tridents at both ends of the shape. These tridents can be explained as follows. Because the locations of the trident shapes are offset from the center

Chapter: Numerical Simulation of MEFISTO

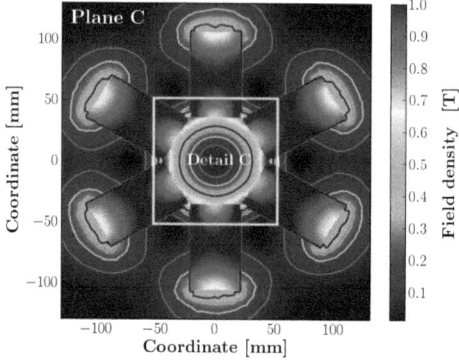

Figure 5.9: Simulated magnetic field density in plane C.

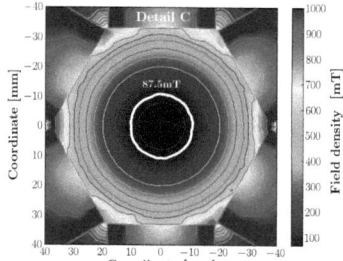

Figure 5.10: Detailed view of the center part of plane C. The ECR isocontour line is highlighted. Toward the center the isocontour lines are concentric circles.

of the magnet arrangement in positive and in negative z direction the radial magnetic field component of the outer ring magnet clusters no longer cancel out. This leads to a dominance of the outwards magnetized cluster in positive z direction. Consequently the radial field of the outward magnetized hexapole magnets is enhanced and the field of the inwards magnetized hexapole magnets is suppressed where the trident shape in positive z direction is located. The trident shape in negative z direction is formed in the same way with the field of the inward magnetized hexapole magnets suppressed and the field of the outwards magnetized hexapole magnets enhanced. Hence, the shapes are twisted relative to each other by 60 degrees (see image to the right in Figure 5.11) because of the same angle separating inward and outward magnetized permanent magnets in the hexapole cluster. The cross section through the mid plane is circular (see Fig. 6.8 for 2D contour plot) and has a diameter of 21.0 mm. The overall length of the shape is 37.3 mm.

Chapter: Numerical Simulation of MEFISTO 75

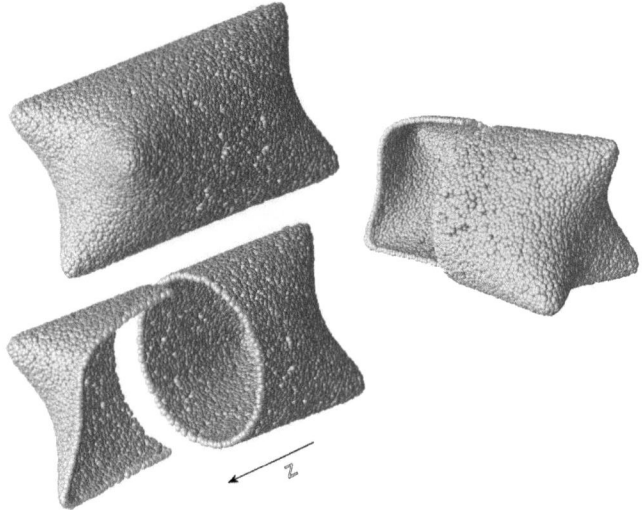

Figure 5.11: Isosurface of a constant magnetic field B_{res} (upper image), cut-away from positive z-direction (lower image) and cut-away seen from negative z-direction (image to the right). The balls at the cutting faces are colored red for better visualization. The triangular shapes at the end of the ECR zone are twisted with respect to each other by 60°.

Cut plane C is located at $z=0$ which results in circular isocontour lines of the field density near the center not similar to the observed trident shaped sputtering patterns observed on diverse hexapole confined ECR ion sources [4]. However if the cut plane is shifted parallel to $z=17$mm the isocontour lines can be identified with the sputtering trident (see Fig. 5.12). This is because the shifted contour plane cuts the isosurface (Fig. 5.11) no longer at the center (as plane C) but 17mm toward the trident face.

We can clearly identify the top face of the isosurface shape with the typical sputtering trident well known for ECR ion sources with a hexapole radial confinement such as MEFISTO. The identification of the simulated and the observed trident takes into account the shape of the ECR region, the presence of high energetic electrons and the subsequent emergence of ions as follows.

According to the ECR plasma model of Wurz et al. ([39]) the ionization is a step by step process starting with few free electrons and a majority of neutrals. The few free electrons get accelerated to high energies by the electric field of the microwave by way of electron cyclotron resonance. The energetic electrons then collide with neutrals to produce singly charged ions and more free electrons. More neutrals get ionized due to the higher electron density. Singly charged ions collide with energetic electrons to be ionized from singly to doubly charged

76   Chapter: Numerical Simulation of MEFISTO

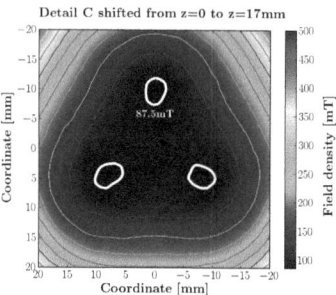

Figure 5.12: Simulated magnetic field density in a plane parallel to C, but shifted by 17mm. While contour plot C at $z=0$ (Fig. 6.8) features circular isocontour lines another contour plot at $z=17$mm reveals the similarity to the sputtering trident. The three separated islands of the ECR isocontour lines are highlighted.

ions and so forth. This way very few free electrons create an avalanche effect and finally establish the quasi stable electron density of the ECR plasma featuring highly charged ions. A fully consistent version of this model has been tested numerically and verified up to a charge state of Ar^{8+} by Hohl et al. [16].

The magnetic trident face has been identified as part of the volume in which ions are created. Ions are also created by the same process in the rest of the depicted shape (Fig. 5.11) as well as inside of it. However only the trident part of the front face is being mapped by the extraction ion optics to the target surfaces where the sputtering has been observed.

The magnetic field density increases in every direction pointing away from the entire depicted shape. The closed depicted shape therefore confines a volume of minimal magnetic flux density. The magnetic arrangement thereby fulfills the criteria of a minimum B-field structure. We shall refer to the defined volume as a plasmoid - a plasma magnetic entity [5]. The presented isosurface element centroids were used to define the launch points of the electron trajectories described in the next section.

5.5 Hot electron trajectories in MEFISTO

A numerical trajectory integration of hot electrons has been performed based on the finite element model of the magnetic field simulation presented in the last section. SCALA, also by Vectorfields, was used to carry out the numerical trajectory integration.

The ECR plasma is slightly negatively charged as described by Shirkov [34, 35] and verified experimentally by Golovanivsky and Melin [14]. Following these measurements the plasma potential has been chosen to be minus 3 V. Due to the compact shape of the simulated plasmoid the spatial distribution of the plasma potential was modeled as a sphere with the same diameter as the plasmoid. The electron heating takes place within the ECR qualified volume sheath with a magnetic field of 87.5 mT including an assumed tolerance of 0.5 mT, which is due to the bandwidth of the implemented microwave generator. Because of collisions and momentum transfer the production of new ions can only happen close to where the electrons are energized. Consequently the initial launch coordinates of the simulated electrons were chosen to be located within the ECR qualified volume sheath. Each bubble seen in Fig. 5.11 is located at an element centroid of the magnetic FEM model and therefore represents one launch point for the trajectory model. In total the isosurface shape yields 25,211 launch points. The ECR heating process mainly takes place perpendicular to the local magnetic field lines [11]. Therefore the initial velocity distribution has been chosen to be anisotropic. The velocity component parallel to the local magnetic field of each trajectory was distributed symmetrically around zero with a Maxwellian temperature of 2 eV [17]. Figure 5.13 shows a histogram of the modulus of this vector component.

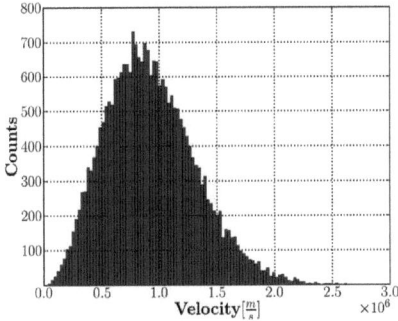

Figure 5.13: The initial velocities of the simulated electron trajectories parallel to the magnetic field are distributed with an according temperature of 2 eV, the temperature of the ions.

In contrast, the modulus of the two velocity components perpendicular to the magnetic field were distributed with a Maxwellian temperature of 2 eV around a kinetic particle energy of 2 keV. Figure 5.14 shows a histogram of the modulus of this distribution.

The model takes into account the magnetic field of the confinement and the electric field of the electron space charge, modeled to be fixed at minus 3 V [14]. It simulates $E \times B$ particle drifts, ∇B and ∇E drifts [7]. The model does not take into account any particle scattering phenomena nor any other kind of particle-particle interaction. The element size in the finite element and the selection criteria of the ECR zone gave a total number of launched particles of 25,211. All simulations have been performed on a IntelCore2Duo 6700 clocked at 2.66 GHz and 2 GB of RAM. The average run time was 83 h at more than 99% of CPU load. Per simulation only one processor core of the Intel2CoreDuo was used.

5.5.1 Trajectory life times and lengths

The trajectory lifetimes, lengths and velocities of each particle were analyzed. Due to hardware limitations the simulations were stopped once less than one percent of the initial number of electrons remained in the simulation. The life

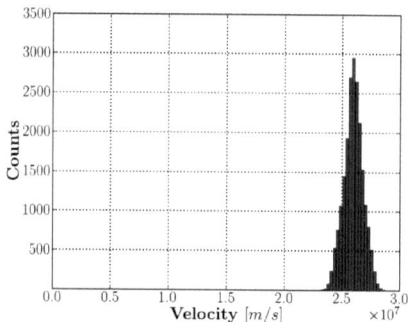

Figure 5.14: The initial velocities of the simulated electron trajectories perpendicular to the magnetic field are distributed with a temperature of 2 eV around a characteristic energy of 2 keV.

time of a particle is limited by the collision with a wall of the plasma container or by the end of the simulation. In a histogram of electron trajectory counts versus trajectory life time we can fit two exponential decay functions of the electron life times τ_1 and τ_2:

$$n(t) = n_0 \left(e^{-\frac{t}{\tau_1}\ln(2)} + \frac{1}{k} e^{-\frac{t}{\tau_2}\ln(2)} \right) \qquad (5.2)$$

Figure 5.15 shows a histogram of the particle life time, an exponential decay with $\tau_1 = 10\mu s$ and another one with $\tau_2 = 36\mu s$ indicated.

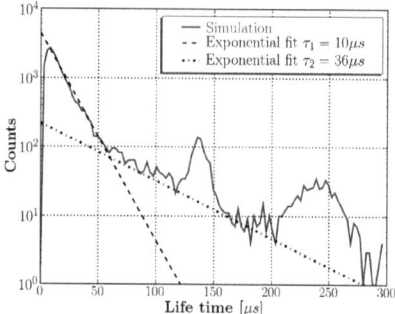

Figure 5.15: Histogram of particle life time and the corresponding exponential fits (simulation A).

There are two significant deviations from the exponential decay behavior around $136\mu s$ and lesser so at $242\mu s$. The deviation represents two electron fractions featuring unexpected long life times. This simulation will be referred to as simulation A. Figure 5.16 gives the mean velocity of each trajectory with respect to its length for this simulation.

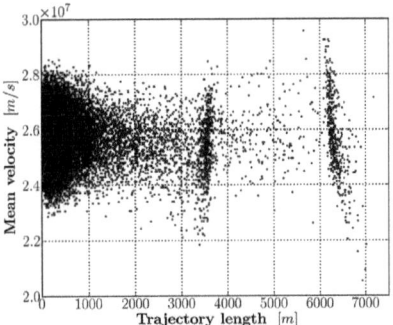

Figure 5.16: Mean velocity versus trajectory length of simulation A. Two clusters at 3537m and 6287m are separated from the main population.

In addition to the main population from 0 to 2000 m there are two clusters of trajectories accumulating at a trajectory length of 3537 m and another one at 6287 m. Both cluster areas present a wide velocity spectrum centered around the same value as the main population. The corresponding trajectories have been traced back to find their underlying launch points. The respective launch points and their initial conditions are distributed homogeneously over both the plasmoid shape and velocity space. Hence, the phenomenon is not the result of a special launch location of the trajectories nor is it due to an anomaly of the initial velocities.

To distinguish the simulated phenomena from numerical errors associated with the random particle velocity distribution a second simulation was performed. The second simulation made use of the same principle of anisotropic initial velocity distribution as the first simulation but starts with a new set of particles. Due to the randomized velocity vector creation process the new set is deliberately different from the first one despite the equal mean energy and the equal initial start locations. Figure 5.17 shows the mean velocity versus trajectory length of the second simulation B.

Table 5.2 gives a statistical breakdown of the trajectory lengths of both simulations.

The high standard deviation indicates the wide spread of the length distribution. The medians of both simulations show a shift of 47.7m. We can see the graphs for both simulation runs are similar in overall shape. Both show the same two clusters in the same area. However in the second simulation the clusters are slightly shifted toward a higher trajectory length which corresponds to the higher median of the second simulation. Table 5.3 summarizes the findings

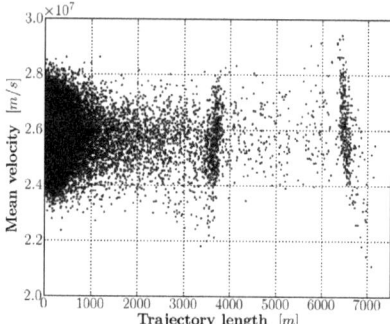

Figure 5.17: Mean velocity versus trajectory length of electrons in second simulation run B.

Simulation	μ	σ
First simulation A	$296.1m$	$1207.6m$
Second simulation B	$343.8m$	$1213.5m$

Table 5.2: A statistical breakdown of the trajectory length of both simulations. μ refers to the median and σ to the standard deviation.

related to both clusters.

Simulation	$\mu_{cluster1}$	$\mu_{cluster2}$
First simulation A	$3537m$	$6287m$
Second simulation B	$3631m$	$6499m$
Δ_{rel}	2.59%	3.26%

Table 5.3: A summary of the clusters separated from the main population. $\mu_{cluster1}$ refers to as the median of the first cluster in each simulation and $\mu_{cluster2}$ to the second one. Δrel depicts the relative difference between each median.

Due to the different initial conditions of velocity distribution the two simulation runs are not identical. Both clusters have a relative shift in length of 2.59% and 3.26%. However both simulation runs show the clusters are clearly separated from the main population and grouped in the same range of trajectory length. Neither median of either simulation corresponds to a dimension given by the plasma container nor a low multiple integer of such.

Electron distribution

In addition to the trajectory length, life times and velocities also their distribution in space can be analyzed. Introducing cut planes parallel to Plane C (see Figure 5.4 we can plot the electron current density resulting from the simulated electron trajectories. Figures 5.18a-d give current density maps in planes spaced 6 mm from each other starting at the plasmoid center and ending at the plasmoid head.

Figures 5.18a-d show, that the electron current density distribution resembles closely the plasmoid shape from a circular cross section at the center of the plasmoid to a triangular shape toward the plasmoid head. The plasmoid and the electron current density are distributed point symmetric with respect to the plasmoid center.

In addition the electron current density distribution maps in Figure 5.18a and 5.18b show that the electrons, which are launched from the ECR surface, do not significantly spread out into the remaining volume outside the ECR surface. However they do expand toward the center of the ECR plasmoid and fill out this volume with a consistently high current density.

In total 25211 trajectories have been launched for a simulation run. The presence of space charge caused by the plasma electrons has been modeled by introducing a charged sphere with the same diameter of the plasmoid and a potential of minus 3 V [17, 5, 14].

5.5.2 Discussion

Moving electrons in a magnetic field are deviated from straight trajectories by the Lorentz force perpendicular to their flight directions. The electrons enter circular orbits with a radius r_L. This is called the Larmour radius of an electron. It depends solely on the electrons momentum and the local magnetic field. It is given by:

$$r_L = \frac{vm}{eB} \quad (5.3)$$

where v is the velocity of the electron, m its mass, e its charge and B is the local magnetic field density. For the case of the MEFISTO ECR zone and 2keV electrons this results in $r_L = 1.722$ mm. The Larmor radii in the ECR zone are therefore much smaller than the plasma container. Hence we can give a rough approximation of revolutions an electron takes before colliding with a container wall with μ as the median of the first simulation run (5.4).

$$n_{rev} \cong \mu/(2\pi r_L) = 2.7 \ast 10^4 \quad (5.4)$$

Given a collisionless plasma the electron motion in the simulated magnetic arrangement is therefore dominated by the magnetic field rather than the container dimensions and could indeed produce long trajectories as it is suggested by the simulation results. However this does not apply for the current operation parameters of the MEFISTO ion source as we will see in Section 5.5.3.

5.5.3 Relevance to laboratory plasmas

Laboratory plasmas as in the MEFISTO facility are not collisionless. The mean free path is limited not only by the plasma container walls but also by collisions with other particles of the plasma and neutrals. According to Chen [7] the mean free path of an electron is given by (5.5):

82 Chapter: Numerical Simulation of MEFISTO

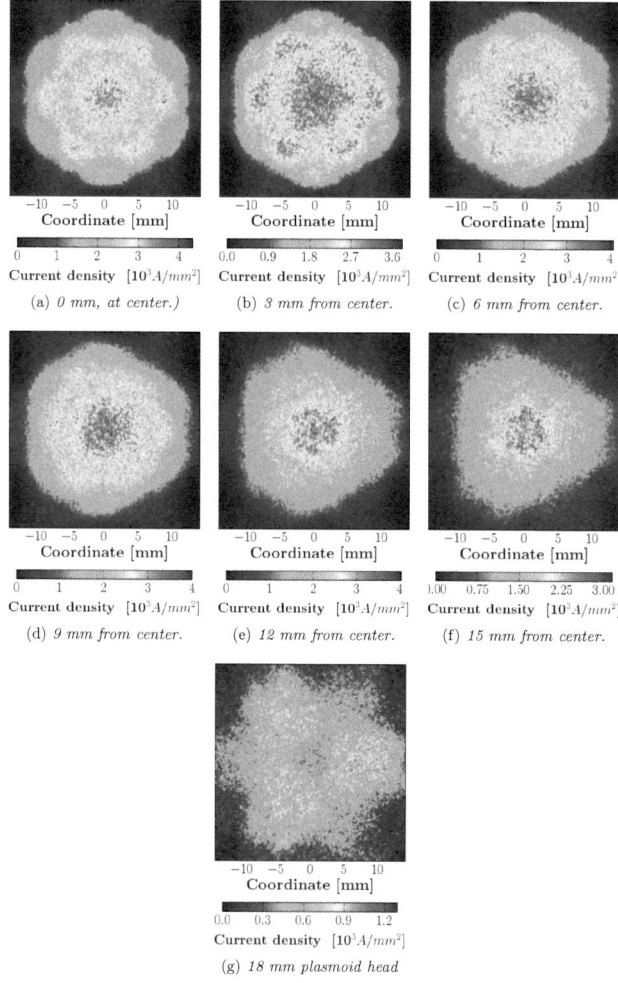

Figure 5.18: Current density maps of MEFISTO in different planes parallel to plane C (see Figure 5.4) spaced 3 mm each.

$$\lambda = 3.4 * 10^{17} \frac{T_{eV}^2}{n \ln \Lambda} \, [m] \qquad (5.5)$$

$$\Lambda = 12\pi n \lambda_D^3 \qquad (5.6)$$

λ_D is the Debye length given by:

$$\lambda_D = \sqrt{\frac{\epsilon_0 K T_e}{n e^2}} \qquad (5.7)$$

where T_{eV} is the electron temperature, K is the Boltzmann constant, n is the plasma density and e is the charge of the electron. Assuming a plasma density of 10^{16} m^{-3} and an electron temperature of 2 keV we get a Debye length $\lambda_D = 3.24$ mm, a Λ of $1.278 * 10^{10}$ and a mean free path length λ of $5.84 * 10^6$ m.

If we assume an effective cross section for electron collisions with neutrals of 10^{-20} m^2 [11], a plasma density of 10^{16} m^{-3} and an ionization fraction of 10% [17] we get a mean free path to the next neutral particle of $\lambda_n = 10^3$ m.

This value shows the mean free path length of 2keV electrons in an ECR laboratory plasma is too short to produce the phenomena suggested by the trajectory simulation. However a plasma density of 10^{15} m^{-3} would lead to a mean free path of 10^4 m. This would bring the simulation results into physical possibility.

5.5.4 Relativistic considerations

Due to the involved electron energies of 2 keV a slight change of electron mass occurs due to special relativity. The electron cyclotron resonance condition depends on the electron mass and the magnetic field density (3.1). A change in electron mass therefore results in a different resonance condition of the qualified magnetic field density. The relation is as follows:

$$\frac{B_1}{B_0} = \frac{m_1}{m_0} = \gamma = \frac{1}{\sqrt{1 - (\frac{v}{c})^2}} \qquad (5.8)$$

v is the velocity of the electron, c the speed of light in vacuum, B_1 the ECR qualified magnetic field density with the consideration of mass increase and B_0 without the consideration of mass increase. v/c equals $8.67*10^{-2}$ and $B_1/B_0 - 1$ equals $3.78 * 10^{-3}$. This modified magnetic field requirement would be fulfilled at a different location due to the spatial distribution of the magnetic field. In fact the isosurface depicted in Section 5.4 would be enlarged toward the plasma chamber walls where the magnetic field features a higher modulus and where the resonance condition would again be fulfilled. However the ECR zone thickness is given by the bandwidth Δf_{MW} of the microwave generator which in turn determines the qualified magnetic field span ΔB_{ECR}. We can write:

$$\frac{\Delta f_{MW}}{f_{MW}} = \frac{\Delta B_{ECR}}{B_{ECR}} \qquad (5.9)$$

The implemented Muegge magnetron [26] features a bandwidth Δf_{MW} of 20 MHz resulting in $\Delta f_{MW}/f_{MW} = 8.1*10^{-3}$. We can state that the ECR zone broadening due to special relativity is smaller than the broadening caused by the bandwidth of the microwave generator:

$$\frac{B_1}{B_0} < \frac{\Delta B_{ECR}}{B_{ECR}} \qquad (5.10)$$

Hence the drift in location of the electron cyclotron resonance toward a higher magnetic field locality due to the effect of special relativity can be neglected.

5.6 Conclusions of the MEFISTO simulation

Future permanent magnet ECR ion sources can be designed very accurately using finite element modeling of the magnetic field. The accuracy of the FEM model was confirmed by comparing it to experimental data from MEFISTO. The location of the ECR qualified plasmoid constrained by an isosurface of constant magnetic field density depends on the magnetic arrangement only. It can therefore be chosen by tuning the design parameters of the magnetic setup.

Confined by permanent magnet hexapoles, long electron life times can be expected for low density plasmas in ECR ion source. This could also be very useful for electron or positron traps or any other particle storage facility as the numerical model is scalable by particle mass, particle energy and magnetic field density. In addition electron current density maps show that the electron distribution is closely related to the magnetic field distribution.

Chapter 6
FEM SWISSCASE

6.1 Introduction

The magnetic field of SWISSCASE has been numerically simulated and experimentally investigated. For the first time the magnetized volume qualified for electron cyclotron resonance at 10.88 GHz and 388.6 mT has been analyzed in highly detailed 3D simulations with unprecedented resolution. The observed pattern of carbon coatings on the source correlates strongly with the electron and the ion distribution in the ECR plasma of SWISSCASE.

There is very little that can be done to change the magnetic field geometry after production of the permanent magnets. It is therefore very important to perform reliable and precise numerical simulations of the magnetic field distribution before the production process to guarantee a successful ECR function.

In this chapter we present the simulation of SWISSCASE and demonstrate the simulation precision that is necessary for the reliable localization of the ECR process. In SWISSCASE the ring shaped magnets establish the axial confinement featuring rotational symmetry with respect to the beam axis. The Halbach hexapole [15] (Figure 4.4b) establishes the radial electron confinement.

The combination of the ring shaped magnets and the hexapole requires a full 3D simulation, rather than two 2D simulations, which would require far less computation power. Therefore, a three dimensional simulation has been performed covering the full spatial extent of the magnetic confinement.

In addition to numerically determining the magnetic field distribution, the simulation also allows calculating electron trajectories. In this paper we show that the resulting electron current density and electron charge density coincide with experimentally observed surface coating and sputtering shapes. Further calculations show the ions are bound to the same location as the high energy electron population. This explains the coincidence of the simulated electron distribution and the observed ion surface coating patterns.

In Section 6.2 the calculation of critical plasma parameters are discussed to justify the results of the numerical simulation. In Section 6.3 the results of the magnetic field simulation are presented. Section 6.4 details the simulated electron distribution followed by the experimentally observed ion coating patterns described in Section 6.5. In Section 6.6 the simulation results are compared to our observations.

6.2 Plasma Parameters

Knowledge of the plasma parameters is required for the calculation of mean free path lengths for high energy electrons as well as low energy ions inside the ECR plasma. It is assumed the plasma is dominated by a cold electron population [17]. It is further assumed the plasma frequency to be the cut off frequency for the incident microwaves inhibiting further microwave plasma heating and limiting the cold plasma electron density [7, 11] (see section 6.1). This leads to a stable equilibrium where the cold plasma density is bound to the microwave frequency.

Our own Bremsstrahlung measurements resulted in a hot electron temperature of 10 keV on average for an argon plasma. Since this kind of measurement and the associated deconvolution methods provide an estimate of magnitude only [3, 12, 19] we can assume the same order of magnitude for the hot electron temperature for a CO_2 plasma (see next section). With a value of 10 keV we obtain the mean free path lengths of the hot electron population according to Table 6.1.

Parameter	Value	Term
$\sigma_{Spitzer}$	$1.21*10^{-23}$ m^2	$(\frac{e^2}{\epsilon_0 E_{kin}})^2 \frac{1}{2\pi} ln\Lambda$
$\lambda_{Spitzer}$	$5.64*10^4$ m	$1/n_e \sigma_{Spitzer}$

Table 6.1: $\sigma_{Spitzer}$: collision cross section for Spitzer collisions, $ln\Lambda$: Coulomb logarithm, n_e: plasma electron density, E_{kin}: kinetic energy of the electrons.

Different formulas for the calculation of mean free path lengths according to different collision mechanisms can be found in the literature (Table 6.2).

Coll. mech.	λ	Ref.
Spitzer	$5.64*10^4$ m	Geller [11]
Single 90°	$1.0*10^6$ m	Chen [7]
Semi empirical	$6.0*10^5$ m	Huba [18]

Table 6.2: Mean free path lengths for different collision mechanisms.

For the source operation with CO_2 gas as will be discussed in the next sections, the following calculation shows the collision mechanism between electrons and neutrals lead to far shorter mean free path lengths than predicted by electron ion collisions. The neutral population consists of CO_2, CO, O_2 and O_3 molecules and of O and C radicals. We simplify the neutral population to a combined total density of $2*n_{CO2}$. It is further assumed this hypothetical population to feature an effective cross section given by a disc with a radius equal to half the length of a CO_2 molecule (116.3 pm). This results in an effective cross section of $\sigma'_{CO2} = 4.25 * 10^{-20}$ m^2. The mean free path can be calculated by Eq. (6.1):

$$\lambda_{CO_2} = \frac{1}{2 n_{CO2} \sigma'_{CO2}} \tag{6.1}$$

λ_{CO_2} is the mean free path in the CO_2 neutral gas background and n_{CO2} its density. Based on fluid and thermodynamical calculations (see Chapter 4) and assuming the neutral gas temperature to be 300 K (see Chapter 3), the neutral gas density in the plasma chamber is $4.805 * 10^{18}$ $1/m^3$ resulting in $\lambda_{CO_2} = 4.90$ m. The limiting mechanism for the mean free path is given by electron collisions with neutrals rather than electron collisions with electrons or ions.

Another effect that must be considered for the validation of this trajectory simulation, is radial electron loss across the magnetic field lines by collision induced diffusion. A simplified expression for the diffusion process of the hot electrons is given by Eq. 6.2 [7].

$$\Gamma_e = -\mu_e\ n_e\ E - D_e\ \nabla n_e \qquad (6.2)$$

Γ_e is the loss rate of the electrons per time and area unit. μ_e is the electron mobility, n_e the electron density, E the electric field and D_e the electron diffusion coefficient. D_e is given by Eq. 6.3 [11].

$$D_e = r_L^2\ \nu_{e-n} \qquad (6.3)$$

r_L is the Larmor radius of the hot electrons and ν_{n-e} the dominating collision frequency between electrons and neutrals as shown in the previous paragraph. With $r_L = 0.868$ mm and $\nu_{e-n} = 1.21 * 10^7$ 1/s we obtain $D_e = 9.13$ m^2/s.

Furthermore μ_e is related to D_e by the Einstein-Smoluchowski equation (Eq. 6.4) [7, 8].

$$\mu_e = \frac{e\ D_e}{K\ T} \qquad (6.4)$$

K the Boltzman constant and T the hot electron temperature. With $T = 10$ keV we obtain $\mu = 9.14 * 10^{-4}$ m^2/Vs. We can now compare the two terms on the right side of Eq. 6.2:

$$\mu_e\ n_e\ E = 3.34 * 10^{17}\ \frac{1}{sm^2} \qquad (6.5)$$

$$D_e\ \nabla n_e = 1.21 * 10^{21}\ \frac{1}{sm^2} \qquad (6.6)$$

E is the electric field assumed to be the plasma potential of minus 3 V divided by the plasma chamber radius $R = 12$ mm. ∇n_e has been estimated by the electron density n_e divided by the plasma chamber radius. From the obtained values we can see the diffusion process is dominated by the term $D_e\ \nabla n_e$ rather than the term $\mu_e\ n_e\ E$. This allows simplifying the calculation of the number of collisions with neutrals an electron undergoes before being lost by diffusion (Eq. 6.7).

$$n_{coll} = T_D\ \nu_{e-n} \qquad (6.7)$$

T_D is the mean time the electrons spend in the confinement before being lost by the discussed diffusion process. ν_{e-n} is the electron collision frequency with neutrals. We can write Eq. 6.7 as Eq. 6.8:

$$n_{coll} = \frac{n_e \, V_{Plasma} \, \nu_{e-n}}{\dot{n}_{De} \, A_{Plasma}} \qquad (6.8)$$

V_{Plasma} is the plasma volume, $\dot{n}_{De}$ the electron loss rate equal to the right hand side of Eq.6.6 and A_{Plasma} the plasma surface. We approximate the geometry of the ECR plasma by a cylinder with length and diameter of the ECR plasmoid. Further introducing $\epsilon_{VA} = V_{Plasma}/A_{Plasma}$ as the ratio of the plasma volume and the plasma surface leads to Eq. 6.9:

$$n_{coll} = \frac{n_e \, \epsilon_{VA} \, \nu_{e-n} \, R}{D_e \, n_e} = \frac{\epsilon_{VA} \, R}{r_L^2} \qquad (6.9)$$

With our values one obtains $n_{coll} = 47.8$. This means hot electrons with a kinetic energy of 10 keV undergo about 48 collisions with neutrals before being expelled from the plasma by diffusion across magnetic field lines. Relating to the CO_2 gas operation discussed further down a similar calculation for ions yields an average of 22 collisions with neutrals before a singly charged carbon ion is lost by diffusion.

We conclude from this section that the calculated mean free paths of the hot electrons are long enough to fit many times into the plasma container and that diffusion across magnetic field lines is not impeding the trajectories of the hot electrons because many collisions are needed to complete a successful diffusion loss mechanism. Hence the trajectory integration based on the simulated magnetic field presented in the next section is valid. Furthermore, considering diffusion across magnetic field lines, ions are confined in the same order of magnitude as electrons in terms of number of collisions before loss.

6.3 The magnetic field

For all magnetic simulations of SWISSCASE, the software TOSCA by *vectorfields* [4], a finite element solver, is used. A summary of the essential parameters of the finite element model are presented in Table (6.3).

Attribute	Quantity
number of linear elements	$4.7 * 10^6$
number of quadratic elements	$4.8 * 10^5$
number of nodes	$1.6 * 10^6$
number of edges	$6.1 * 10^6$
RMS	1.033
distortion of worst element	$1.40 * 10^{-3}$

Table 6.3: Some characteristics of the finite element model used for the numerical calculation of the magnetic field.

Figure 6.1 shows the finite element model used for the magnetic field simulation. Cut planes have been introduced for the visualization of section views.

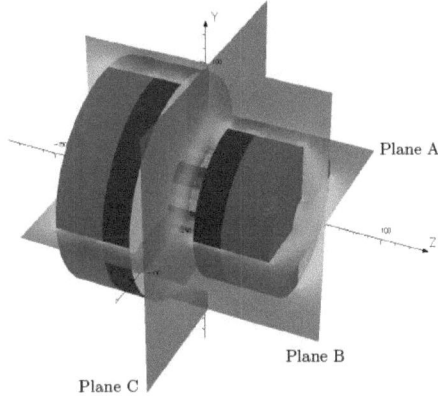

Figure 6.1: Cut planes are introduced to better visualize the magnetic field presented in Figures 6.3 to 6.8. All magnetic parts of the finite element model have been meshed carefully according to size and relevance thereby optimizing the simulation efficiency.

The axial magnetic field has been measured using a Hall probe. The simulation result did not deviate from the measurement more than 2.8 % at any point along the $z - axis$ (see Figure 6.1). Figure 6.2 shows an overlay of the simulated and the measured magnetic field.

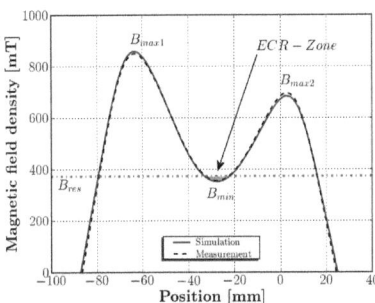

Figure 6.2: Overlay of measured (dashed black line) and simulated (blue line) axial magnetic field for comparison. The maximal error is 2.8% at the maxima to the right, toward the extraction.

The comparison shows that the simulated and the measured magnetic field are in good agreement along the z-axis. To investigate the full 3D extent of the simulated magnetic field section cuts are shown, followed by a 3D representation of the plasmoid, defined by an isocontour surface.

Figure 6.3 to Figure 6.8 give section views of the magnetic model defined by the cut planes introduced in Figure 6.1 and a detailed view of the central part of each plane respectively.

As expected from the non symmetric ring magnet assembly, Figure 6.3 reveals the non symmetric magnetic field distribution. The asymmetric field distribution due to the asymmetric ring magnets improves the extraction process because the decreased magnetic mirror ratio on the extraction side (to the right in Figure 6.4) leads to a weaker ion confinement and hence a better ion extraction.

Figure 6.4 shows, the 388 mT isocontour line shown in Figure 6.4 is slightly distorted and asymmetric. This is due to the superposition of the hexapole field and the ring magnetic field. Despite the pronounced asymmetry of the ring magnets the field distribution inside the hexapole is rather symmetric with respect to cut plane C (see Figure 6.1).

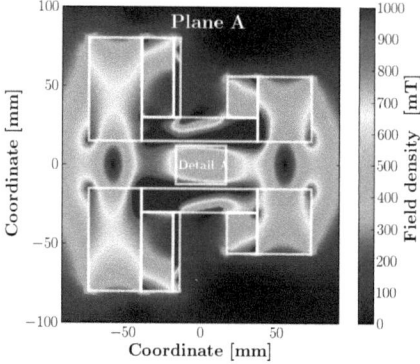

Figure 6.3: Simulated magnetic field in plane A.

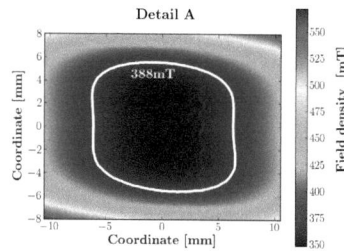

Figure 6.4: Detailed view of the center part of plane A. The ECR isocontour line is highlighted, revealing the topography of the ECR qualified surface.

In Figure 6.6 is shown that the same contour line is symmetric in a cut plane B, oriented 90° relative to plane A (see Figure 6.1).

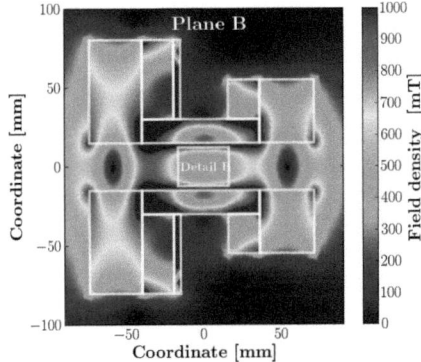

Figure 6.5: Simulated magnetic field in plane B.

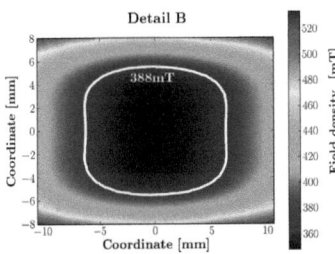

Figure 6.6: Detailed view of the center part of plane B. The ECR isocontour line is highlighted. Cut plane B is rotated by 90 degrees with respect to plane A. The contour lines are symmetric with respect to the beam axis.

The magnetic field distribution in plane C (see Figure 6.8) appears circular at the plasmoid center and close to the 388 mT isocontour line. However the magnetic field is no longer circular at locations further outside the plasmoid, dominated by the hexapole confinement.

The ECR qualified surface, the plasmoid, presents itself compact and point symmetrically deformed as expected from the permanent magnet setup. Figure 6.6 shows, despite its larger size than the permanent magnetic ring to the right (positive z-coordinate), the permanent magnet ring to the left (negative z-coordinate) does not deform the isocontour line in any asymmetric way. However the difference in size of the larger permanent magnet ring shifts the center of the plasmoid along the z axis toward the larger permanent magnetic ring (in Figure 6.4 to the left) without distorting its overall shape on a larger scale.

For a better visualization Figure 6.9 shows the plasmoid composed of spheres centered at each finite element qualified for the ECR isosurface. The plasmoid

Chapter: FEM SWISSCASE

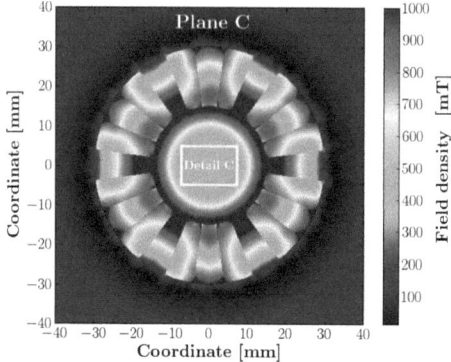

Figure 6.7: Simulated magnetic field in plane C.

Figure 6.8: Detailed view of the center part of plane C. The ECR isocontour line is highlighted. Toward the center the isocontour lines are concentric circles.

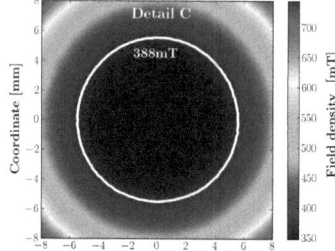

surface was cut and the cut edges colored red to better reveal its 3D shape.

The simulation results show the design of the ECR zone location is a rather delicate procedure. A slightly different magnetic field setup, such as an asymmetry in the permanent magnet assembly, leads to significant deformation and delocalization of the ECR zone. The rather thin ECR zone defined by the bandwidth of the microwave generator (see Chapter 3 for ECR theory and Chapter 5 for ECR zone broadening) defines the maximum finite element size in the ECR region. Larger elements in this region lead to strong discretization errors and to a massive loss of simulation precision. The modeling of SWISSCASE showed that the elements needed to precisely model the ECR zone with a 1% error margin, are dominated by the remaining elements which model the permanent magnets, the soft iron rings and the background. Hence saving computation power on the ECR zone is not economic.

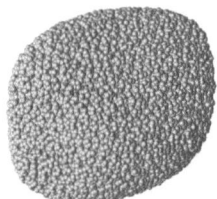 

Figure 6.9: SWISSCASE plasmoid (left image) and cut way (right image) composed of ECR isosurface qualified spheres. For better visualization the spheres along the plasmoid cut edges are shaded red.

6.4 The electron distribution

In addition to the solution of the magnetic model, electron trajectories have been calculated using the finite element model trajectory integrator [4] SCALA by *vectorfields*. The initial velocities of the particle trajectories were chosen to be anisotropically distributed. The electron velocity vector components perpendicular to the local magnetic field lines are distributed with a temperature of 10 keV, representing the hot electron temperature which was determined by the Bremsstrahlung measurement presented in Chapter 8. The velocity vector components parallel to the local magnetic field are distributed with a temperature of 2 eV [17], the temperature of the cold electrons. This anisotropic velocity distribution complies with the ECR heating mechanism with the same anisotropic orientation which is perpendicular to the local magnetic field lines [7]. A total of 36'328 trajectories have been launched for a simulation run. The presence of space charge caused by the plasma electrons has been modeled by introducing a charged sphere with the same diameter of the plasmoid and a potential of minus 3 V [14, 17, 5].

The magnetic field distribution of the MEFISTO ECR ion source located at the University of Bern has been simulated by Bodendorfer et al. [4] and matches the observed surface coating triangles in MEFISTO. The simulated magnetic field distribution of SWISSCASE does not resemble the respective surface coating triangles in SWISSCASE to the same extent. However, a closer look at the electron distribution instead of the magnetic field distribution, suggests a strong link between electron distribution and surface coatings in SWISSCASE. Figures 6.10a-f illustrate how the shape of the simulated electron current density distribution changes from a hexagon at the plasmoid center into a triangular shape at the tip of the microwave antenna, where a triangular-shaped surface coating has been observed (Figure 6.11) after the operation of the SWISSCASE ECR ion source [4]. This transformation of the electron density maps represents the changing cross section of the ECR zone while the cut plane is moved from the plasmoid center toward the antenna. Note, due to the geometry of the magnetic field, the plasmoid and the electron density distribution are point symmetric with respect to the plasmoid center.

In lack of our own Bremsstrahlung measurement at the time, an earlier simulation run has been performed with an electron energy of 25 keV (see Ap-

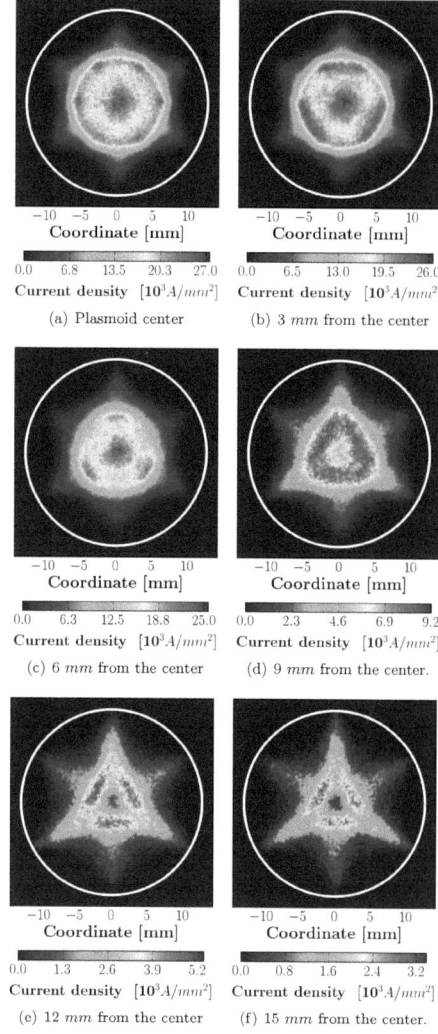

Figure 6.10: Electron current density maps at different positions along the optical axis. Note the increasing similarity with surface coating triangles (see Figure 6.11) toward the antenna tip at 15 mm offset from the plasmoid center. The white ring represents the plasma chamber wall (diameter 24 mm). Dimensions in mm. Simulated electron energy: 10 keV.

pendix **??**). The electron current density patterns are very similar, differing in magnitude rather than shape. The Bremsstrahlung measurement presented in Chapter 8 does not exclude the presence of an additional electron population with a mean energy of 25 keV which has been measured by R. Friedlein et al. in a 7.5 GHz ECR ion source [10].

In the next section, the observed coating pattern and a spectral analysis is presented. It shows the coating consists indeed of particles found in the plasma.

6.5 The observed surface coating pattern

Figure 6.11 shows a typical surface coating triangle resulting from the exposure directly or indirectly to the ECR plasma. In this particular case the observed surface coating triangle was at the tip of the microwave antenna (see Figure 4.4a) in the SWISSCASE facility after the operation of a CO_2 plasma. In some cases the observed pattern is a sputtered area rather than a surface coating. This may originate from the difference in particle energies the surface is exposed to. The observed surface coatings have been exposed directly to the plasma without any kind of extraction or post acceleration of the plasma particles. In this case the particle energies may have been low enough to enable a deposition process of the neutralized carbon atoms to form the observed surface coating. On the other hand sputtering shapes are observed after the extraction and post acceleration where the involved particles have energies many orders of magnitude larger than they had inside the plasma. In this case the particle energy might be too large for a continuous deposition process and hence sputtering occurs.

The antenna tip was chemically analyzed to determine the composition of the surface coating. The result is summarized in Table 6.4.

Element	Weight [%]	Abundance [%]
Carbon	90.61	93.04
Oxygen	7.36	5.67
Fluorine	1.41	0.91
Neon	0.62	0.38
Total	100	100

Table 6.4: Composition of the surface coating in mass and abundance fraction.

The dominance of carbon in the coated surface suggests a deposition process originating from the plasma because the plasma mainly consists of carbon and oxygen neutrals and ions. Oxygen is found as a minor fraction only because it mainly recombines with other oxygen and carbon radicals and molecules to form molecular oxygen, ozone, carbon monoxide or carbon dioxide. All of these recombination products are gaseous and therefore not likely candidates for a deposition process. FC40, a liquid composed of Fluorine and Carbon, N-$(C_3F_7)_3$, is used as cooling fluid enveloping the plasma chamber. Fluorine combines well with metals. Therefore, the Fluorine traces in the spectral analysis may originate from cooling fluid diffusing in small amounts through the copper plasma chamber wall. Neon found in the spectral analysis is a remainder of previous Neon plasma operations in the same plasma chamber.

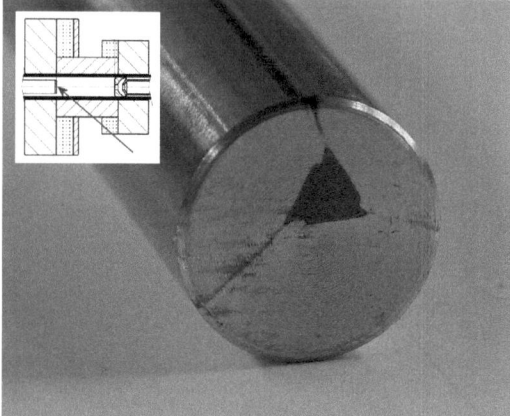

Figure 6.11: Photograph of the microwave antenna of SWISSCASE after operation (CO_2, 100 W input power, 10.48 GHz), published by Bodendorfer at al. in [4]. A dark triangular patch consisting mostly of carbon is clearly visible, which was not present before the operation. The shape of the deposited coating matches the electron density map, as explained in the text and in Figure 6.10. The pattern is displaced from the optical axis due to a misalignment of the microwave antenna caused by the antenna's length and weight. The misalignment did not cause any measurable decrease in the ion beam performance. The antenna has a diameter of 12 mm and the side length of the triangle measures 4.2 mm.

6.6 Relation between simulation and observation

In this section we explain the relation between the observed surface coating pattern and the simulated electron distribution.

As discussed in Section 6.2 the electron mean free path is limited to 4.9 m by collisions with neutrals. The mean free path is therefore far longer than the ECR plasmoid. This in turn potentially allows ionization processes anywhere inside the plasma container. However the magnetic simulation and the numerical trajectory integration show that the electrons are confined to a limited space inside the confining field (see Figures 6.10a-f for the electron density profile).

To investigate the spatial distribution of the ions we make the approximation that the ion-neutral cross section equals to twice the value of the hot electron-ion scattering cross section ($8.5*10^{-20}$ m^2) and a neutral density equal to the value presented in Section 2 of this chapter ($4.805*10^{18}$ $1/m^3$). This leads to a mean free path length for ions of 0.8 m which is 62 times the length of the ECR plasmoid. This suggests the ions would completely fill all of the vessel volume resulting in an uniform density distribution. However in a magnetic field of 388.6 mT singly charged carbon ions with a kinetic energy of 2 eV gyrate with a Larmor radius of 1.82 mm around the magnetic field lines on which they were created. The trajectories of the carbon ions are therefore restricted by the magnetic confinement field rather than the limited geometry of the plasma vessel. Hence the ions do indeed travel for a long mean free path length relative to the ECR plasmoid dimensions. But the ions are prevented by the magnetic field from filling all the vessel volume and are confined to a restricted space similar to the electron distribution presented in the last section.

This explains the good agreement between the observed surface coating triangle and the simulated electron distribution because the carbon ions do not travel far from the region of ionization defined by the presence of high energy electrons.

Relativistic considerations

Due to the electron energies involved of 10 keV a slight change of electron mass occurs due to special relativity. The electron cyclotron resonance condition depends on the electron mass and the magnetic field density (3.1). A change in electron mass therefore results in a different resonance condition of the qualified magnetic field density. The relation is as follows:

$$\frac{B_1}{B_0} = \frac{m_1}{m_0} = \gamma = \frac{1}{\sqrt{1-(\frac{v}{c})^2}} \qquad (6.10)$$

Where v is the velocity of the electron, c the speed of light in vacuum, B_1 the ECR qualified magnetic field density with the consideration of mass increase and B_0 without the consideration of mass increase. v/c equals $19.77*10^{-2}$ and B_1/B_0-1 equals $2.01*10^{-2}$. This modified magnetic field requirement is fulfilled at a different location due to the spatial distribution of the magnetic field. In fact the isosurface depicted in *Section 6.3* would be enlarged toward the plasma chamber walls where the magnetic field features a higher modulus and where the resonance condition would again be fulfilled. However the ECR zone thickness is given by the bandwidth of the open loop microwave generator which in turn determines the qualified magnetic field span as described in Chapter 5. In addition the magnetic simulation error is 0.97% and the permanent magnetic material is susceptible to room temperature. These three effects summarized, the ECR zone thickness due to the microwave generators bandwidth (0.45%), the simulation error (0.97%) and a room temperature change of 5 K (0.75%), result in a bluring of the ECR zone by 2.17% which is larger than the effect induced by special relativity (2.01%).

6.7 Conclusion

We simulated in detail the 3D magnetic field distribution of the SWISSCASE ECR ion source. The simulation of high energy electron trajectories based on the same magnetic model leads to shapes nearly identical to the observed ion surface coating triangle. This implies the ion distribution is related to the high energy electron distribution. This is justified by the magnetic field parameters, the ion mean free path lengths and similar Larmor radii and diffusion parameters of hot electrons and singly charged carbon ions in SWISSCASE.

Furthermore we generalize that lightweight ions with Larmor radii and mean free path lengths similar to those of a hot electron population can be considered as spatially bound to the latter. This can significantly simplify the calculation of the ion distribution in ECR related applications such as ECR ion engines and ECR ion implanters.

Chapter 7

Source characterization and performance

In this chapter, the characterization and performance of SWISSCASE are presented. Firstly, an effect is explained which requires re-adjustment of the microwave antenna for each new microwave frequency setting. The effect is detailed and the cause for it revealed. Ion mass spectra of SWISSCASE represent the central section of this chapter, characterizing the ion output of SWISSCASE operated with different gases. Argon, krypton, xenon and carbon dioxide spectra are presented and explained. Finally, the power input required for the ionization of the extracted ions is calculated, based on the presented spectra and the ion beam composition. To estimate the overall efficiency of SWISSCASE, the calculated ionization power is compared to the actual microwave power which is radiated into the ECR plasma.

7.1 Microwave coupling

The ECR plasma can be operated at different microwave frequencies. The frequency available from the microwave generator is in the range from 8.5 to 11.5 GHz. However, as will be mentioned further down, not the whole frequency band is of interest with respect to the extracted ion beam performance. In the following we will show the frequency dependent ion current output of SWISSCASE in a restricted frequency band which yields an interesting performance.

7.1.1 Frequency dependence

From Chapter 4 the minimum of axial magnetic field inside the SWISSCASE confinement is 356.3 mT. This magnetic field value corresponds with an electron cyclotron frequency of 10.0166 GHz (see Chapter 3 for details). Below this frequency there is no corresponding ECR zone. Hence no ECR effect is supposed to occur. During the experiment a sharp decrease in ion current was found once this frequency was undercut. However useful ion current with little noise and low effervescence were found only for frequencies higher than 10.48 GHz.

Extracted ion current of Ar^{8+}

To give a useful overview of the frequency dependence, the ion beam performance with respect to one particular charge state has been recorded. The chosen charge state is Ar^{8+} because this charge state can be found in many 10 GHz ECR ion sources [11, 26, 16] and shows the ability of the system to produce high charge states. Each data point represents optimized settings for all other input parameters such as feed gas pressure, position of the short, puller electrode position and ion optics voltages. Figure 7.1 shows the maximum extracted currents of Ar^{8+} ions versus the applied microwave frequency.

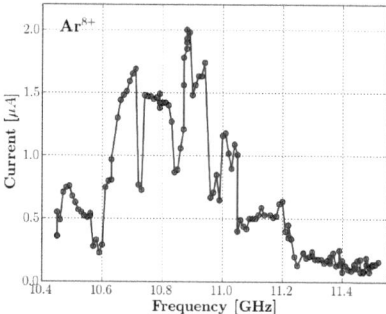

Figure 7.1: Extracted ion current of Ar^{8+} versus applied microwave frequency. Microwave power was found to be ideal at maximal power settings throughout the hole frequency band at is thus not strong enough. Note the steps in the ion current which are explained further down this section.

From Figure 7.1 we see the optimal microwave frequency is 10.88 GHz leading to a maximum extracted ion current of 2 μA. The measurements start at a frequency of 10.48 GHz because unstable and monotonous decreasing low currents were observed for lower frequencies. The measurements are limited to 11.52 GHz by the frequency limit of the microwave generator (see Chapter 4. The ion current shows significant fluctuations throughout the applied frequency band. However, a broad band of multiple maxima is observed between 10.6 and 11.1 GHz. In the following section we will go into more detail about the microwave antenna position settings applied to obtain the data points shown in Figure 7.1.

Antenna position

The installed microwave power generator is able to deliver linearly polarized microwaves with a frequency of 9 to 11.5 GHz. The power output of the microwave generator depends on the frequency as pointed out in Section 4. The wave guide system including the circulator, both microwave windows and the UHV system also feature a frequency dependent transmission curve. Finally the plasma itself is sensitive to the frequency of the incident microwave because the frequency defines the location of the ECR region and the different plasma cut off densities. We will show that despite these numerous factors, one particular parameter, namely the microwave antenna position, dominates the transmission efficiency.

As the frequency of the microwave is changed, the antenna position and the position of the short have to be adjusted to maintain the impedance matching of the system. Experience showed the optimal position of the short depends on the antenna position. This means for each antenna position there is a single optimal position of the short. Both positions, short and antenna, are adjusted iteratively to find the optimal position. Figure 7.2 shows the necessary antenna position for each frequency setting in Figure 7.1.

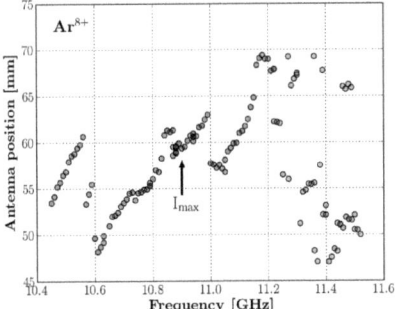

Figure 7.2: Microwave antenna position versus applied microwave frequency. Optimal operation points up to 11.22 GHz in blue appear coherent (explained further down this section) while data points beyond this frequency in green are scattered. The black arrow indicates the frequency which yields the maximum extracted ion current as seen in Figure 7.1.

In Figure 7.2 we can see jumps in the otherwise smooth data below 11 GHz. When the microwave frequency is increased to a point where a jump can be seen in the diagram, no more increase in ion current is observed and another, lower, value for the antenna position has to be chosen to continue on a high performance output combination. The new antenna position correlates with the wavelength of the applied microwaves. To better understand this we introduce Figure 7.3.

Correlating the distance between the antenna tip and the ECR plasma would result in an inversely decreasing antenna position with increasing frequency which leads to hyperbolas in Figure 7.2. This is because the wave-length of the injected microwave decreases with increasing frequency. This would not lead to the observed pattern of increasing antenna position with increasing frequency. However another concept of investigation leads to the effective antenna length to fulfill geometric conditions to act as an efficient transmitter. Depending on the frequency the length of the antenna has to allow a resonant condition according to classical antenna theory.

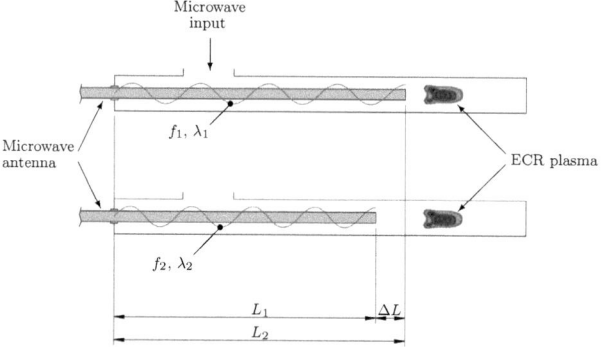

Figure 7.3: Schematics of microwave situation for two different microwave frequencies and adjusted antenna position. We are interested in the effective antenna lengths L_1 and L_2 rather than the distance between the antenna tip and the ECR plasma.

For simplicity we consider a TE01 microwave mode in the plasma chamber. This requires the last fraction of the microwave to be $\pm \lambda_1/4$. Combining the requirements we can write down the following relation (Eq. 7.1).

$$L_1 = n\frac{\lambda_1}{2} \pm \frac{\lambda_1}{4} \tag{7.1}$$

L_1 is the effective antenna length in the first case with microwave frequency f_1. Due to symmetry we only consider $L_1 = n\lambda_1/2 + \lambda_1/4$ and neglect $L_1 = n\lambda_1/2 - \lambda_1/4$. In the second case we apply a different microwave frequency f_2, for instance $f_2 < f_1$. Analogue to Eq. 7.1 we can write Eq. 7.2:

$$L_2 = n\frac{\lambda_2}{2} + \frac{\lambda_2}{4} \tag{7.2}$$

L_2 is the effective antenna length in the second case with microwave frequency f_2. We can further write the necessary adjustment ΔL to the antenna position as follows Eq. 7.3:

$$\Delta L = L_1 - L_2 = n\frac{\lambda_1}{2} + \frac{\lambda_1}{4} - n\frac{\lambda_2}{2} - \frac{\lambda_2}{4} \tag{7.3}$$

Introducing $\lambda = c/f$ with c as the speed of light and f as the microwave frequency we can simplify Eq. 7.3 to calculate the expected adjustment of the antenna position:

$$\Delta L = (\frac{n}{2} + \frac{1}{4})(\frac{f_2 - f_1}{f_1 f_2})c \tag{7.4}$$

From Eq. 7.1 we can see that n is the number of half wavelengths fitting into the effective antenna length at the optimal antenna position of frequency f_1. In our case the start of our measurements defines this frequency to be 10.45 GHz, the corresponding wavelength of 28.69 mm and the optimal effective antenna length of 328.3 mm. Hence n = Integer(2*328.3 mm / 28.69 mm) = 23. n is conserved throughout the calculation. ΔL is recalculated for each frequency. Figure 7.4 visualizes the expected antenna adjustments by a red line, according to Eq. 7.4 using n = 23. Additionally dashed red lines indicate the expected antenna position in case subsequent phase shifts of $\frac{\lambda_2}{4}$ are subtracted from the actual calculated position.

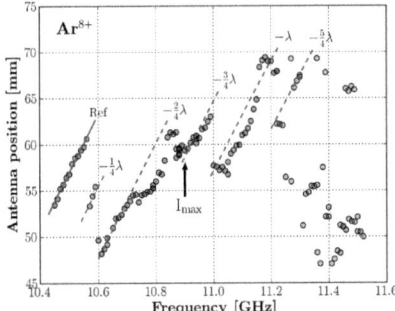

Figure 7.4: Circles are measured optimum microwave antenna positions versus applied microwave frequency (same data as Figure 7.2). Red lines indicates the expected adjustment of the antenna position by Eq. 7.4. Dashed red lines show the same position corrected by subsequent phase shifts of $\frac{\lambda}{4}$.

In Figure 7.4 we can see the measured points indeed follow the prediction (Eq. 7.4) up to a frequency of 10.56 GHz where the first jump occurs with a phase shift of $\frac{\lambda}{4}$. A similar behavior can be observed at 10.59 GHz. However from a frequency of 10.7 GHz the measured points fit the predicted positions worse. A better fit is achieved again around 10.93 GHz. Despite that the general tendency toward the predicted position is still visible, the fit gets worse with increasing frequency and is barely possible to correlate beyond 11.2 GHz.

7.1.2 Conclusion of the frequency dependence

The extracted ion output shows a significant variation with the frequency of the microwave input. For different frequencies the coupling between microwave generator and plasma has to be maintained by adjusting the effective antenna length. The necessary change in the effective antenna length is dominated by the transmission criteria of the antenna rather than the distance between antenna tip and ECR plasma. This suggests the space between the antenna tip and the ECR plasma does not allow a proper formation of a microwave radiation field. This is supported by the fact that this distance is in the range of 36 to 13 mm, very close to the wavelength of a 10.88 GHz microwave (27.55 mm). Hence we are in a near-field situation rather than a far-field situation. In this case the electric field of the microwave is significantly distorted and modified by the presence of the plasma chamber walls and the ECR plasma. A closer investigation of the electric field distribution would require numerical simulations involving the detailed geometry of the plasma chamber and a sophisticated microwave plasma model. Both are beyond the scope of this thesis.

7.2 Ion beam performace

7.2.1 Measurement setup

The second Einzel-lens is a remainder of the original experimental ECR setup from the University of Giessen, where Trassl et al. [36] performed numerous tests with and without this separate Einzel-lens. In Giessen this Einzel-lens improved the performance of the 10 GHz multi-mode ion source, tested in 1999, for some ion species of Bismuth (see [36]).

Despite the high transmission ratio of the ion optics and the mass separation magnet described in chapter 4, experiments were performed to test the ion beam performance without the second Einzel-lens in place. At the same mass per charge resolution the removal of the second small Einzel-lens yielded an ion beam current improvement between 5% and 15% for all tested operation gases. Consequently all presented spectra were produced without the second Einzel lense in place.

To analyze the extracted ions the ion beam is separated into a spectrum by the mass separation magnet discussed in more detail in Section 4. After separation the remaining ion beam is collected by a Faraday cup and neutralized. The ions hitting the Faraday cup capture electrons from the Faraday cup surface and create a positive potential difference. The potential difference in the Faraday cup is allowed to equilibrate by a ground connection. The resulting current necessary to maintain the charge equilibrium of the Faraday cup is measured by a Keithsley 6517 pico ampere meter. Figure 7.5 shows a schematic of the measurement setup.

The magnetic field B splits the extracted ion beam into different mass-per-charge ions. The magnetic field is measured by a Bell Digital Gaussmeter 811AB with a Bell x10 lateral hall sensor. The data of the hall sensor and the Faraday cup are stored by a National Instruments PXI 1042Q data logger and stored in a PC with LabView to compose the mass spectra presented in this section. In addition to the connections shown in Figure 7.5, the current driver for the mass separation magnet is controlled by a computer. The the microwave input power and frequency settings, antenna and position of the short, all pressure sensor outputs and the cooling fluid temperature are stored by the data logger. Table 7.1 gives a summary of the settings and read-outs.

The preference on manual settings in most of the operation parameters of SWISSCASE in the experimental phase guaranteed a quick access and allowed the development of an operational feeling for fine tuning. The optional computer setting allows to control all input parameters by computer except the antenna position and the position of the short. However additional automation would also allow to control the antenna and the position of the short to enable full

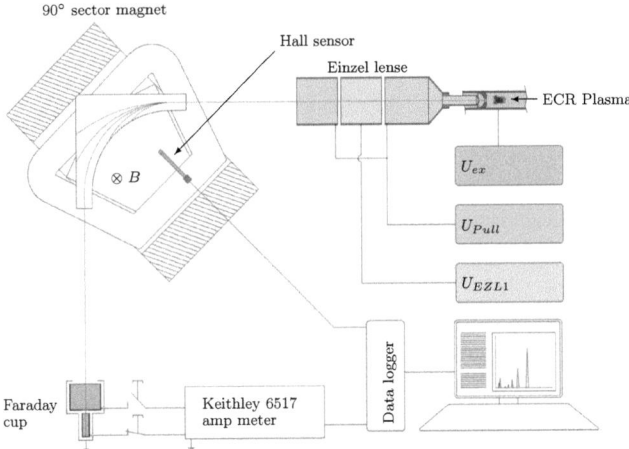

Figure 7.5: Schematic of measurement setup used to determine the extracted ion current of SWISSCASE. For the presented spectra: $U_{ex} = 10$ kV, $U_{Pull} = 0$ kV, $U_{EZL1} = 7.2...8.8$ kV. The second Einzel-lens presented in Chapter 4 is removed for improved ion beam performance.

automatic operation.

7.2.2 Faraday cup

Louville's theorem [11] states the emittance integral of an ion beam cannot be changed by electro-static or magneto-static fields. Unless baffles and wall collisions lead to loss related reshaping of the beam we consider the extracted ion beam emittance integral as conserved throughout the whole beam transfer system from the plasma meniscus to the Faraday cup.

Ion source parameters optimized for maximal ion current output of a specific charge state are not the same as for a source setup optimized for best charge state resolution. They differ in the the high voltage settings for the first Einzel-lens and the puller electrode position rather than the plasma parameters. Test runs performed by Broetz, Trassl et al. [6, 36] showed that maximal ion current output is achieved at the expense of charge state resolution. The lower charge state resolution is caused by a wider beam leading to wider charge state peak widths. This effect originates in the lower space charge density in a wider beam. This is optimal for high current extraction and low beam divergence.

Hence a larger aperture is required to allow a wider ion beam being captured by the Faraday cup than for the narrower ion beam resulting from a high resolution setup.

Table 7.1: Parameter settings and logging of SWISSCASE. M: manual setting preference, (M): manual optional, C: computer logging preference, (C): computer optional, (V): visual readout optional. P_1: Pressure in target chamber, P_2: Pressure in ion optics cube, P_3: Pressure at feed gas inlet port, U_{ex}: Extraction potential, U_{EZL1}: first Einzel-lense voltage and U_{Pull}: puller electrode voltage (see figure 7.5).

Parameter	Setting	Logging
Microwave power	M (C)	C (V)
Microwave frequency	M (C)	C (V)
Antenna position	M	C (V)
Position of the short	M	C (V)
Feed gas valve	M (C)	C (V)
P_1	-	C (V)
P_2	-	C (V)
P_3	-	C (V)
U_{ex}	M (C)	C (V)
U_{EZL1}	M (C)	C (V)
U_{Pull}	M (C)	C (V)
Mass separation	C (M)	C (V)
Ion current	-	C (V)

Faraday cup design

To capture both modes, high current and high resolution, we designed a new Faraday cup with two separate stages featuring two different apertures. The large aperture of 50 mm allows to capture a wide ion beam in high current mode. The smaller aperture of 10 mm allows to analyze an ion beam in high resolution mode with an aperture area 1/25th of the area of the large aperture. This principle has been realized in a two stage concept where the small aperture is situated behind the large aperture. In high current mode operation both apertures are connected in parallel. In high resolution mode the large aperture is grounded and the small aperture only is used for measurements. Figure 7.6 and 7.7 show a section cut and a 3d view of the new designed Faraday cup used for all ion current measurements presented in this thesis.

The design of two different apertures operated simultaneously enables us to switch between apertures purely electrically with no moving parts and without breaking the vacuum.

Secondary electron suppression

The ion beam hitting the Faraday cup creates secondary electrons leaving the surface with a typical kinetic energy of up to about 45 eV. This effect increases the measured current in addition to the ion current if the secondary electrons manage to leave the Faraday cup. Hence specific measures had to be taken to suppress the secondary electrons leaving the Faraday cup.

Figure 7.6 shows that a sophisticated rotational symmetric sawtooth profile has been realized at the bottom of the Faraday cup to further decrease the rate of secondary electrons leaving the Faraday cup.

The usage of a gridded Faraday cup was not chosen because of additional complications due to interactions between the grid and the ion beam. Instead an electrostatic field emitted from two baffles was used to create a potential distribution along the beam axis which prohibits secondary electrons from leaving the cup.

A finite element model has been created to design and optimize the potential distribution necessary for successful secondary electron suppression. Figure 7.8

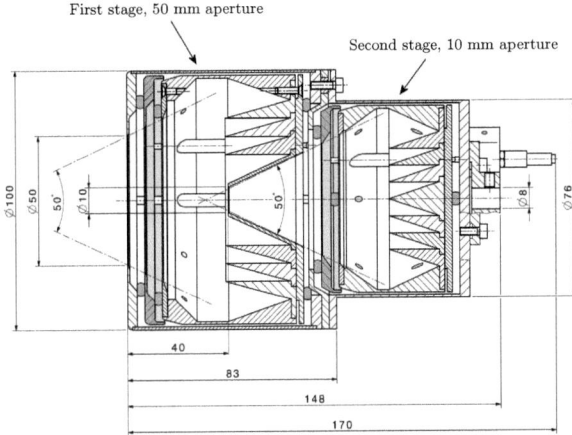

Figure 7.6: Section cut of the two stage Faraday cup. Cyan colored are the two secondary electron suppression baffles. Brown colored are the synthetic ruby balls providing the necessary insulation. The large aperture is used in high current mode where a larger beam is present. The smaller aperture is used in high resolution mode with a narrower beam, where the ion beam passes through the first stage and is eventually collected in the second stage.

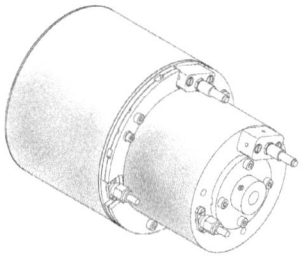

Figure 7.7: 3D view of the newly designed two stage Faraday cup.

shows the optimized potential distribution and a 3D view of the Faraday cup.

Figure 7.8a shows the tail of the potential seen by the incident ions is less than 5 V in magnitude at a position of 50 mm before the Faraday cup entry aperture. This potential tail leads to a focusing effect on the incident ion beam. However compared to the kinetic energy of 10 keV per ion charge, the potential tail is considered to be negligible.

Test electrons were launched in the finite element model to simulate the secondary electron emission. Figure 7.9 shows a cut visualization of the FEM

Figure 7.8: Potential distribution. Baffle voltage: -200 V. A section cut is superimposed for better orientation.

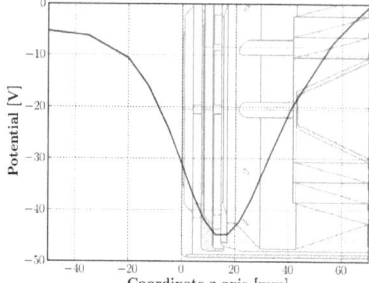

model.

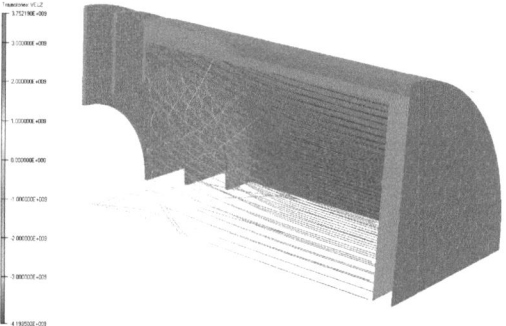

Figure 7.9: One quarter of the finite element model with trajectories used to determine the optimal geometry and potentials for successful secondary electron suppression. The color bar indicates the electron velocities in mm/s.

Figure 7.9 shows electron trajectories with an initial launch energy of 45 eV. With a baffle voltage of -200 V secondary electrons with a kinetic energy equal or less than 45 eV are trapped by the potential well.

Faraday cup conclusion

In practical application the secondary electron suppression proved to be essential for accurate current measurements. With the suppression voltage switched off the measurements indicate an ion current up to 20 % larger compared to measurements with the suppression voltage switched on. Also the magnitude of the suppression voltage of minus 200 V has been veryfied by the observation that any further increase in voltage did not lead to a further decrease in the ion current measurement readings.

We conclude from this section that the design, simulation and operation of the newly realized Faraday cup with two different apertures is a full success.

7.2.3 Argon spectra

In this section the performance of SWISSCASE to produce different charge states of argon will be presented. For all argon ECR plasma operation and argon spectra, argon gas of natural isotope composition has been used. Table 7.2 gives an overview of the natural abundance of the different argon isotopes [37].

Argon isotope	Natural abundance [%]
^{36}Ar	0.337
^{38}Ar	0.063
^{40}Ar	99.60

Table 7.2: Natural abundance of stable argon isotopes. ^{40}Ar is dominant.

During the acquisition of the presented spectra none of the mentioned isotopes besides ^{40}Ar have been identified.

Argon spectra low and medium charge states

As ion current receiver the staged Faraday cup is used, which was presented in more detail in Section 7.2.2. For best resolution which is necessary for the higher charge states of argon the small aperture with a diameter of 10 mm is used, rather than the large aperture with a diameter of 50 mm. Figure 7.10 shows a spectrum of an argon ECR plasma extraction. The ion source parameters are optimized for the charge state of Ar^{6+} in this case to bring out the medium charge state performance of SWISSCASE.

The Ar^{6+} output peaks at 6.01 μA. The source parameters are optimized for Ar^{6+} and are suboptimal for all other charge states such as Ar^{8+}. This can be seen by comparing the Ar^{8+} output of this setup and the setup optimized for Ar^{8+}. The ion current of Ar^{8+} in the Ar^{6+} optimized setup peaks at 1.75 μA. However the ion current of Ar^{8+} in the Ar^{8+} optimized setup peaks at 2 μA (see Figure 7.1). The difference comes from a different feed gas pressure setting and different optimal voltages of the ion optics. The plasma state optimized for Ar^{8+} requires a lower feed gas pressure than the plasma optimized for Ar^{6+}. Since ionization is a step by step process [16] between each ionization step there is a certain probability for charge exchange with neutrals or lower charged ions. On the other hand neutrals and lower charged ions are the base where highly charged ions are derived from. This defines an optimal equilibrium neutral density where as many neutrals as necessary are present to sustain the ionization chain but as few neutrals as possible are present to impede high charge states by charge exchange and recombination. This equilibrium neutral density is lower for higher charge states hence the lower feed gas pressure for a plasma optimized for Ar^{8+} rather than for Ar^{6+}.

For these charge states of argon the current peaks are spaced wide enough to use the large aperture of the Faraday cup with a diameter of 50 mm. Figure 7.11 shows a spectrum acquired with the large aperture active.

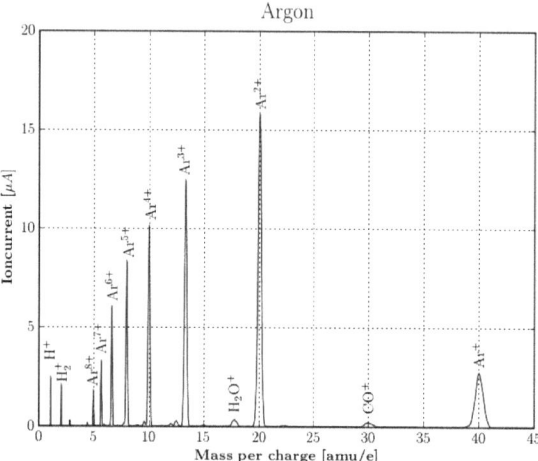

Figure 7.10: Spectrum of argon plasma of SWISSCASE. The ion source parameters have been optimized for the charge state of Ar^{6+}. Plasma parameters: $f = 10.88$ GHz, $P = 85$ W (P_{max}), $p_3 = 6.6 * 10^{-4}$ mbar.

Figure 7.11 shows that the large aperture collects significantly more ions than the small aperture. However the full width half maximum (FWHM) peak widths acquired with the large aperture are also larger than the FWHM peak widths of the small aperture. Table 7.3 summarizes the peak currents, the gain ratios and the ratio of the FWHM widths for each charge state in this setup:

Table 7.3 shows that the higher gain in ion current also brings the deficit of a higher FWHM width which directly impedes resolution. The FWHM for both the 10 mm and the 50 mm aperture can be slightly improved. However this always is at the expense of the ion current gain.

In addition to argon also H_2O^+ and CO^+ are present as residuals from the last venting procedure and from the previous carbon dioxide plasma operation, respectively.

Argon spectra, high charge states

In this section the performance of SWISSCASE to produce highly charged argon ions is presented. In order to resolve the current peaks of higher charged ions for all subsequent spectra the 10 mm aperture of the Faraday cup has been used. Figure 7.12 gives a spectrum of an argon ECR plasma optimized for Ar^{12+}.

In Figure 7.12 we see different charge states of argon from Ar^{9+} to Ar^{12+}. We cannot see a possible peak of Ar^{13+} because it is concealed by the peak of C^{4+}. Ar^{14+} has not been detected. The flat tops of Ar^{10+}, O^{4+}, C^{3+}, Ar^{9+} and

Chapter: Source characterization and performance   111

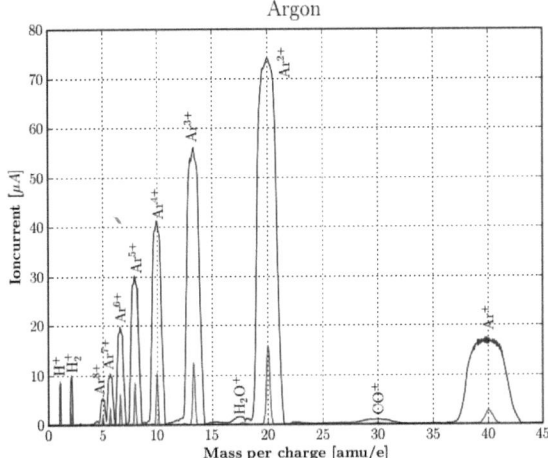

Figure 7.11: Spectrum of the same argon plasma as shown in Figure 7.10 with the same plasma parameters in blue. For comparison the original spectrum acquired with the small aperture is shown in red.

Charge state	current 10 mm aperture	current 50 mm aperture	gain ratio I_{50}/I_{10}	FWHM ratio W_{50}/W_{10}
Ar^+	2.73 μA	16.9 μA	6.19	4.75
Ar^{2+}	15.8 μA	73.9 μA	4.68	4.50
Ar^{3+}	12.4 μA	55.8 μA	4.50	6.00
Ar^{4+}	10.1 μA	41.3 μA	4.09	3.91
Ar^{5+}	8.33 μA	30.0 μA	3.60	4.93
Ar^{6+}	6.01 μA	19.7 μA	3.28	5.08
Ar^{7+}	3.24 μA	10.3 μA	3.18	5.89
Ar^{8+}	1.75 μA	5.50 μA	3.14	5.11

Table 7.3: Summary peak currents acquired with the small aperture of 10 mm diameter and the large aperture of 50 mm diameter. I_{50}/I_{10} gives the ratio between the two currents and W_{50}/W_{10} is ratio of the FWHM peak widths at the same charge state.

N^{3+} are due to the range setting of the Keithley pico ampere meter limited to 2.5 nA.

In addition to argon we can also see contamination with species of ions such as oxygen, carbon and nitrogen. The oxygen and the carbon ions are residuals

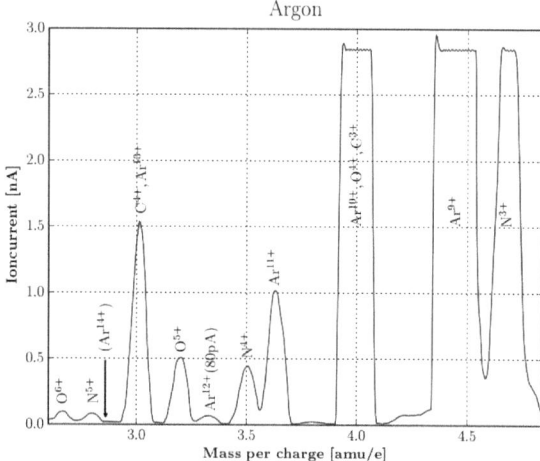

Figure 7.12: Spectrum of an argon plasma optimized for Ar^{12+}. Plasma parameters: $f = 10.88$ GHz, $P = 85$ W (P_{max}), $p_3 = 1.2 * 10^{-4}$ mbar.

from a former CO_2 plasma operation. The nitrogen is a residual from the last atmospheric venting. Note the close coincidence occurring with Ar^{10+}, O^{4+} and C^{3+} in the peak at $m/q = 4.0$ and the coincidence occurring with C^{4+} and Ar^{13+} in the peak at $m/q = 3.0$ and $m/q = 3.09$, respectively. For all argon plasma parameter settings the most abundant charge state is Ar^{2+}.

Argon performance conclusion

We conclude from this section that all expected charge states of argon are represented in the acquired spectra well above noise level and measurement accuracy. The maximum charge state which has been measured is Ar^{12+} peaking at 30 pA followed by Ar^{11+} (1 nA), Ar^{10+} (16 nA) and Ar^{9+} (80 nA). The current reading of Ar^{10+} includes possible contamination with O^{4+} and C^{3+} ions.

7.2.4 Krypton spectra

In this section the performance of SWISSCASE to produce different charge states of krypton will be presented. For all krypton ECR plasma operation and krypton spectra, krypton gas of natural isotope composition has been used. Table 7.4 gives an overview of the natural abundance of the different krypton isotopes [37].

Compared to the natural isotope composition of argon, krypton features an isotope composition which is spread out over more isotopes. As we will see the

Krypton isotope	Natural abundance [%]
^{78}Kr	0.35
^{80}Kr	2.25
^{82}Kr	11.6
^{83}Kr	11.5
^{84}Kr	57
^{86}Kr	17.3

Table 7.4: Natural abundance of stable krypton isotopes.

wider isotope distribution of krypton results in wider peaks and even separate peaks of the same charge state in a current versus mass per charge spectrum.

Krypton performance, low and medium charges states

Figure 7.13 gives an overview of a typical krypton spectrum optimized for Kr^{8+}.

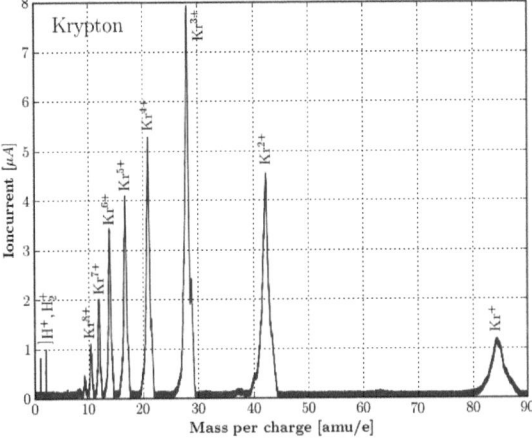

Figure 7.13: Spectrum of a krypton plasma optimized for Kr^{8+}. Plasma parameters: f =10.63 GHz, P = 95 W (P_{max}), p_3 =3.1*10^{-4} mbar.

In the overview of the krypton spectrum we see different charge states present from Kr$^+$ up to Kr^{8+}. This shows SWISSCASE is able to provide a krypton spectrum as expected from its performance with respect to argon. However the krypton spectrum is different from the argon spectrum insofar that the different isotopes in their natural abundances in the krypton spectrum are visible. Figure 7.14 provides a magnified view of the two krypton charge states Kr^{2+} and Kr^{3+}

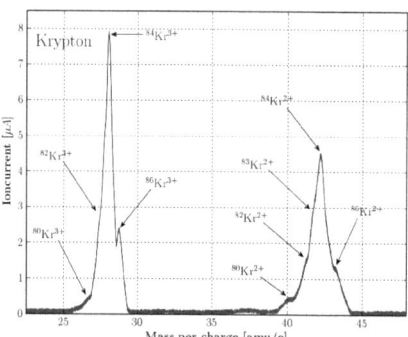

Figure 7.14: Magnified view of the peaks Kr^{2+} and Kr^{3+} in the spectrum shown in Figure 7.13. In both peaks different isotopes can be recognized. In addition to the main peaks of $^{84}Kr^{2+}$ and $^{84}Kr^{3+}$, we can also identify several other isotopes such as $^{86}Kr^{2+}$ and $^{86}Kr^{3+}$.

from the same spectrum shown in Figure 7.13 revealing different isotopes in the substructure of each ion current peaks.

The magnified view of the krypton spectrum presented in Figure 7.14 features ion current peaks significantly broadened by the natural isotope composition of the krypton gas used for the ECR plasma. To investigate the high charge state performance of SWISSCASE, narrow peaks with little or no isotope broadening are required. However SWISSCASE is able to produce high charge states of krypton. Because of the higher atomic mass of krypton compared to argon we can separate higher charge states from each other. This is due to the lower rate of coincidences with other ions with the same mass per charge ratio.

Krypton performance, high charge states

Figure 7.15 shows a krypton spectrum of SWISSCASE optimized for Kr^{13+}. In addition to krypton ions there are also ions of oxygen and carbon as remainder of a previous carbon dioxide plasma operation. C^{2+} conceals the potential occurrence of Kr^{14+}. Hence the highest identified charge state is Kr^{13+} (3 nA). For all krypton plasma parameter settings the most abundant charge state is Kr^{3+}.

Krypton performance conclusion

We conclude from this section that SWISSCASE is able to produce all charge states of ionized krypton up to Kr^{13+}. The isotope distribution given by the natural abundance in krypton leads to wider peaks compared to the argon spectra. Despite the broader peaks, high charge states of krypton can still be identified due to the larger atomic mass of krypton gas and consecutive lower rate of coincidences with other ions with the same mass per charge ratio.

7.2.5 Xenon spectra

Xenon is the gas with the highest atomic mass which SWISSCASE has been tested with in the framework of this thesis. As for spectra of argon and krypton, also for xenon the natural isotope composition has been used to operate the ECR

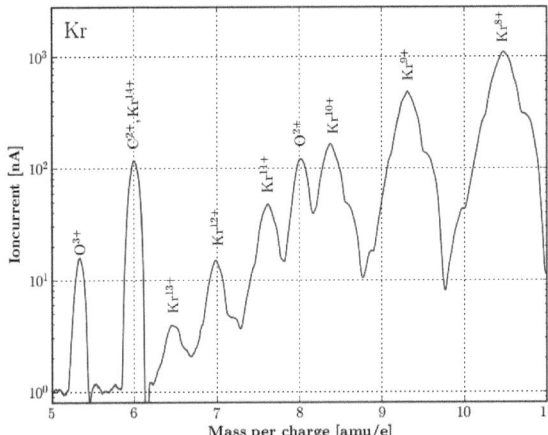

Figure 7.15: Krypton spectrum optimized Kr^{13+}. The mass per charge resolution is limited by the broad isotope distribution leading to relatively wide krypton peaks. Plasma parameters: f =10.63 GHz, P = 95 W (P_{max}), p_3 =2.3*10^{-4} mbar.

plasma. Table 7.5 gives an overview of the natural abundance of the different xenon isotopes [37].

Compared to the natural abundances of krypton isotopes, xenon isotopes are distributed more broadly with the highest abundance of ^{132}Xe at 26.9 % and with more stable isotopes than krypton. As we will see this leads to even wider peaks.

Xenon spectra, low and medium charge states

Figure 7.16 gives an overview of a typical xenon spectrum obtained from SWISS-CASE.

In the spectrum presented in Figure 7.16 we can see different charge states of xenon represented by single peaks. Unlike in the krypton spectrum we cannot resolve distinct peaks for xenon isotopes. This comes from the close distribution of the isotope abundances of natural xenon leaving no gapes between isotopes large enough for observation with our mass separation system.

Xenon spectra, high charge states

Despite the wider peaks we can resolve high charge states of xenon. Figure 7.17 shows a cut out of the spectrum presented in Figure 7.16.

116 Chapter: Source characterization and performance

Table 7.5: Natural abundance of stable xenon isotopes.

Xenon isotope	Natural abundance [%]
^{124}Xe	0.1
^{126}Xe	0.09
^{128}Xe	1.91
^{129}Xe	26.4
^{130}Xe	4.1
^{131}Xe	21.2
^{132}Xe	26.9
^{134}Xe	10.4
^{136}Xe	8.9

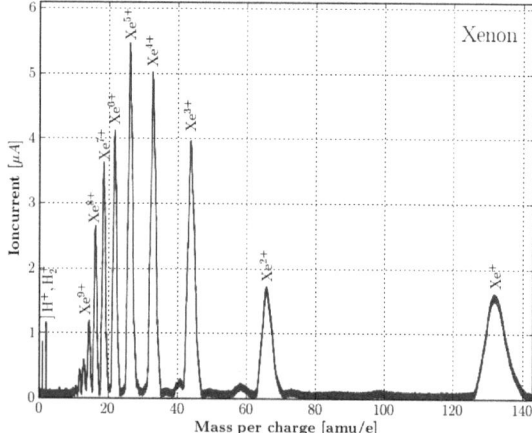

Figure 7.16: Spectrum of an xenon plasma optimized for Xe^{6+}. Plasma parameters: $f = 10.63$ GHz, $P = 85$ W (P_{max}), $p_3 = 1.0 * 10^{-3}$ mbar.

The focus on the higher xenon charge states shown in Figure 7.17 demonstrates the ability of SWISSCASE to produce highly charged xenon ion currents significantly above noise level. The interference of Xe^{11+} and C^+ conceals the true ion current of Xe^{11+}. However the presence of Xe^{12+} indicates the ion current of Xe^{11+} is significant. The coincidence of Xe^{13+} and Ar^{4+} similarly conceals the true Xe^{13+} current. No higher charge states of xenon are visible in the spectrum. The highest identified xenon charge states is therefore Xe^{12+}.

For most xenon plasma parameter settings the most abundant charge state is Xe^{5+}. Only for the highest feed gas pressures required for optimal Xe^{1+} production, the most abundant charge state is Xe^{4+}.

Chapter: Source characterization and performance 117

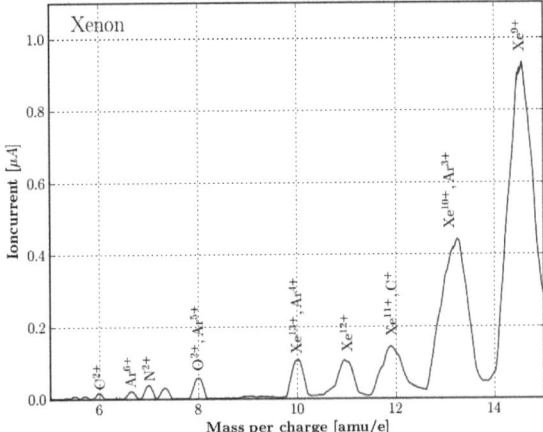

Figure 7.17: Xenon spectrum optimized for Xe^{12+}. Plasma parameters: $f = 10.63$ GHz, $P = 85$ W (P_{max}), $p_3 = 1.0*10^{-3}$ mbar.

Xenon performance conclusion

We conclude from this section that SWISSCASE is able to produce xenon ions of high charge states. The highest charge state identified is Xe^{12+}. Unlike the krypton spectrum no separate isotope peaks are visible in the xenon spectrum due to the limited resolution of the mass separation system and the distribution of the natural xenon isotope abundances.

7.2.6 Carbon dioxide spectra

Carbon dioxide features ions of two different families, oxygen and carbon, both of highest interest for space borne instrumentation. Oxygen is the third most abundant element in the solar wind with charge states up to O^{6+}. Carbon is the fourth most abundant element in the solar wind and can be found in all possible charge states, up to C^{6+}.

Beside the trace contamination of molecular nitrogen present in all SWISS-CASE spectra, Carbon dioxide is the only molecular gas SWISSCASE has been deliberately operated on. Its low chemical reactivity and the absence of any molecular polarity makes it a very practical ultra high vacuum operation gas for SWISSCASE. Carbon dioxide with natural isotope composition has been used. Table 7.6 gives an overview of the natural abundance of the different oxygen and carbon isotopes [37].

In addition to the shown abundances of the stable carbon isotopes there is

Table 7.6: Natural abundance of stable oxygen and carbon isotopes. ^{13}C accounts for about 1% in the natural adundance of carbon. ^{14}C is not listed because it is unstable but needs to be considered anyway due to its half life of 5730 years.

Isotope	Natural abundance [%]
^{16}O	99.76
^{17}O	0.4
^{18}O	0.200
^{12}C	98.90
^{13}C	1.10

^{14}C with a half life of 5730 years which also has to be considered. All other unstable oxygen and carbon isotopes are of no relevance due to their short half lives.

Carbon dioxide spectra, low charge states and overview

Figure 7.18 gives an overview of a typical CO_2 spectrum obtained from SWISS-CASE. The parameters were optimized for C^{3+}.

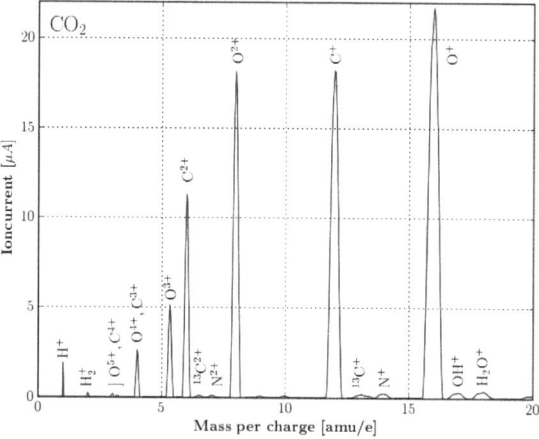

Figure 7.18: Carbon dioxide spectrum optimized for C^{3+}. Plasma parameters: $f = 10.88$ GHz, $P = 85$ W, $p_3 = 6.8 \cdot 10^{-4}$ mbar. Note the $^{13}C^+$ and $^{13}C^{2+}$ peaks.

The carbon dioxide spectrum given in Figure 7.18 gives an expected charge state distribution up to O^{5+}. In this setting the optimal microwave power input is below the maximal possible input power. Higher input power leads to a decrease in ion current of C^{3+}. This effect may be caused by a shift of the ionization equilibrium toward higher charge states with higher input power (see Chapter 3 for more details).

In addition peaks of the stable carbon isotope ^{13}C are visible with charge states 1+ and 2+. This is an expected result since ^{13}C accounts for 1% in the natural carbon abundance (see Table 7.6).

Carbon dioxide spectra, high charge states

Figure 7.19 shows a carbon dioxide spectrum optimized for C^{5+}. All expected charges states are present up to O^{7+}. The spectrum is not shown for $m/q=2$ because the presence of H_2^+ conceals all other ion species with the same mass per charge ratio such as C^{6+} and O^{8+} is therefore not informative. The shown spectrum also indicates the presence of N^{5+} as a trace contamination of N_2 from the last venting procedure.

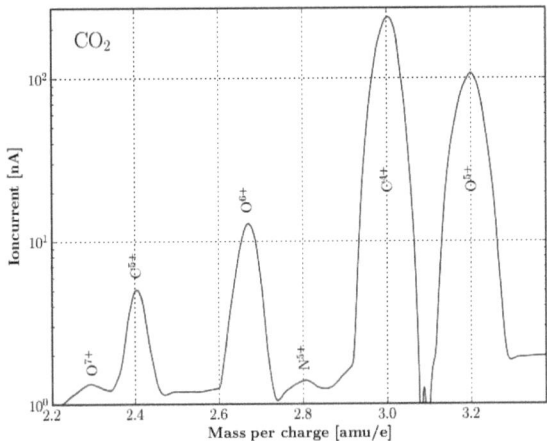

Figure 7.19: Carbon dioxide spectrum optimized for C^{5+}. Plasma parameters: $f = 10.88$ GHz, $P = 90$ W, $p_3 = 5.0 \cdot 10^{-4}$ mbar.

Carbon dioxide performance conclusion

We conclude from this section that SWISSCASE is able to produce oxygen and carbon ions of high charge states. The highest charge states possible to identify are O^{7+} and C^{5+}. O^{8+} and C^{6+} cannot be identified due to an interference in m/q with H_2^+. Isotope separation is not visible due to the low natural isotope abundances and the limited resolution of the mass separation system.

7.3 Required ionization power

7.3.1 Introduction

Ionization is a step by step process evolving many different charge states for highly charged ions. Ions leaving the ECR plasma, caused by the extraction of ions and the formation of an ion beam, represent an energy loss for the ECR plasma. This energy loss is compensated by the microwave input which heats up ECR electrons and insures the resupply of fresh ions by electron impact ionization (see Chapter 3).

Hence there is an energy flow from the microwave, through the ECR electrons, onto the ions and out of the extraction aperture, carried away by the ions. This energy flow is the only inherently necessary energy flow of the ECR ion source. All other energy flows such as electron and ion loss, radiation and heat transfer to the walls are parasitic effects, desired to be as little as possible for best microwave efficiency.

Given an ion spectrum, the total ionization power represented by the extracted ion beam and the respective ionization energy, can be calculated. The resulting extracted ionization power, contained in the ion beam, can be related to the microwave input power to obtain a gross value for the microwave power conversion efficiency.

7.3.2 Beam ionization power

We first calculate the energy necessary for each ionization step of argon ions, convolute the ionization energy with the measured ion abundance found in the mass-spectrum and calculate the ion flux in the ion beam. This results in the extracted ionization power from the plasma.

If E_q is the total energy required to produce the ion, ΔE_k the energy required to increase the charge state of an ion from $k-1$ to k, then the energy required to produce an ion of charge state q is given by Eq. 7.5:

$$E_q = \sum_{k=1}^{q} \Delta E_k \quad (7.5)$$

The number $\dot{n}_q$ of extracted ions of charge state q is given by the respective ion current measured in the Faraday cup (e is the electrons charge):

$$\dot{n}_q = \frac{I}{e\,q} \quad (7.6)$$

With Eq. 7.7 we can calculate the power P_q which was necessary to create the ion beam with current I_q of the respective charges state q. $E_{q[eV]}$ and $E_{q[J]}$ is the total ionization energy, measured in electron volts $[eV]$ and in Jouls $[J]$:

$$P_q = \dot{n}_q \cdot E_q = \frac{I_q}{e\,q} \cdot E_{q[eV]} = \frac{I}{q} \cdot E_{q[J]} \quad (7.7)$$

Performing these calculations for a given spectrum of argon (see Chapter 8) results in table 7.7.

Charge state	ΔE_k [eV]	E_q [eV]	I_q [μA]	$\dot{n}_q$ [1/s]	P_q [μW]
1	15.76	15.759	2.73	$1.70 \cdot 10^{13}$	43
2	27.63	43.388	15.8	$4.93 \cdot 10^{13}$	343
3	40.74	84.128	12.4	$2.58 \cdot 10^{13}$	348
4	59.81	143.938	10.1	$1.58 \cdot 10^{13}$	363
5	75.02	218.958	8.33	$1.04 \cdot 10^{13}$	365
6	91.00	309.965	6.01	$6.25 \cdot 10^{12}$	310
7	124.32	434.284	3.24	$2.89 \cdot 10^{12}$	201
8	143.456	577.74	1.75	$1.37 \cdot 10^{12}$	126
9	422.44	1000.18	0.08	$5.55 \cdot 10^{10}$	8.89
10	478.68	1478.86	0.016	$9.99 \cdot 10^{9}$	2.37
11	538.95	2017.81	0.001	$5.67 \cdot 10^{8}$	0.183
12	618.24	2636.05	$3 \cdot 10^{-5}$	$1.56 \cdot 10^{7}$	$6.59 \cdot 10^{-3}$

Table 7.7: Argon plasma. Summary of step wise ionization energy ΔE_k, total ionization energy E_q, measured ion current I_q, resulting particle flux $\dot{n}_q$ and ionization power P_q. The total power, summarized over all shown charge states, results in 2111 μW. Other ion species in spectrum, such as H$^+$ or N$^+$ are not taken into account.

From Table 7.7 it follows that the ionization power required to deliver the measured ion beams is a small fraction compared to microwave generator output power of 95 W. Figure 7.20 gives a diagram

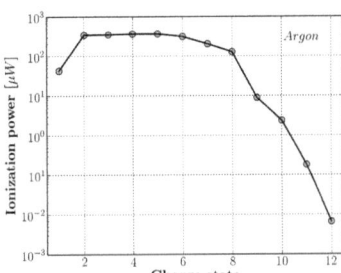

Figure 7.20: Extracted ionization power for each charge state of argon from Ar$^+$ to Ar^{12+}. The ionization power for each charge state ranges from $6.59 \cdot 10^{-3}$ μW to 365 μW featuring a power plateau between Ar^{2+} to Ar^{8+}.

From Figure 7.20 it can be seen that the ionization power features a plateau from Ar$^+$ to Ar^{8+}. This suggests a close relation between available ionization power provided by the ECR electrons and the resulting charge states. However for charge states higher than Ar^{8+} the curve steadily declines revealing a more complex relation.

7.3.3 Discussion

The above calculation neglects ions and ECR electrons which are lost from the plasma and delivers a power estimation purely based on the extracted ions. However the extracted ions are a good representation of the plasma ion composition [16, 11].

The total power, summarized over all calculated charges states (2.1 mW, see caption in Table 7.7) is a small fraction of the total microwave generator output power. The effective microwave power, absorbed by the ECR plasma is not accessible at this point. This is due to the complex and highly dynamic absorption and reflection property of any ECR plasma [11]. To reveal the microwave power absorbed by the ECR plasma, numerical simulations, including a microwave plasma model, could be used (see Chapter 9) in conjunction of a bi-directional microwave coupler measuring ingoing and reflected power at a certain point in the transmission line.

Despite the reputation of ECR ion sources to be very energy efficient in the production of highly charged ions [11], the overall efficiency is still quite modest. For SWISSCASE the ratio between microwave generator output power and necessary ionization power based on the measured ion beam results in: 2.1mW/95W = 0.021 %. Hence there is still a huge potential for efficiency improvement.

The majority of the microwave power is either reflected from the ECR plasma, absorbed in the dummy load or in the facility walls. Both, the dummy load and parts of the facility become warm and even hot to the touch, suggesting that significant microwave power is lost or reflected during the transmission and the coupling process.

No further investigation was performed regarding absorbed microwave power in the ECR plasma.

7.4 Summary of characterization

SWISSCASE is able to produce very useful currents of highly charged ions of argon, krypton, xenon, oxygen and carbon. It can be assumed that its ability for high charge state production extends to other gaseous elements, metals and composites. From the charge state distribution found from extracted ion beams of argon, krypton and xenon plasmas, the most abundant charge states are multiply charged ions rather than singly charged ions. The most abundant charge states do not change significantly with feed gas pressure. Only in the xenon plasma the most abundant charge state shifts from Xe^{5+} to Xe^{4+} with increasing feed gas pressure.

The ion source showed very reliable and reproducible plasma ignition behavior for all tested operation gases. Ignition feed gas pressure was found to be close to the gas pressure optimal for the lowest charge states.

The ion output of each charge state can be optimized individually by optimizing feed gas pressure, Einzel-lens voltage, microwave input power and frequency. However, the ratio of non optimized and fully optimized charge state does not exceed a factor of 5 generally. This leads to the comfortable situation where all charge states except the highest ones are present in a wide range of plasma parameters. This enables robust and quick parameter finding of higher ion charge states beginning from parameter settings optimized for low charge states.

The optimal microwave power setting for singly and doubly charged ions of argon, krypton and xenon were 85W, slightly below the maximum setting. However, for all charge states higher than two and for all charge states of carbon and oxygen, the optimal power setting was at its maximum. This suggests that a further increase in input power would also increase the ion output.

The optimal microwave input frequency lay within 10.63 GHz and 10.88 GHz for all tested operation gases. The gases with more massive nuclei, xenon and krypton, are operated more efficiently at 10.63 GHz rather than 10.88 GHz, which turned out to be optimal for argon and carbon dioxide. This difference in microwave input frequency leads to a shift of the ECR zone inside the magnetic confinement producing a larger ECR volume for the heavier ions of xenon and krypton which seems to be beneficial for the respective ion beam current. An explanation for this effect has not been found yet.

Comparing the microwave power consumption with the necessary ionization power based on the extracted ion beam composition, SWISSCASE shows a moderate energy efficiency. The transmission, absorption and conversion of microwave energy into ionization energy is a very complex process, which needs a numerical plasma and facility model in order to further investigate any relation between microwave power input and ion charge state output.

Chapter 8

Bremsstrahlung measurement

In this chapter the results of a Bremsstrahlung measurement are presented. The Chapter is divided into an introduction presenting the target of observation, hot ECR electrons. The measurement setup is introduced in a dedicated section, presenting the used X-ray detector and explaining the associated calibration process of it. In Section 8.3 the results of the Bremsstrahlung measurement are checked for their validity and are critically discussed. Eventually a restricted energy region of the obtained Bremsstrahlung spectrum is selected to determine the temperature of the hot ECR electrons.

8.1 Introduction Bremsstrahlung measurement

Calculations of the ion temperature based on emittance measurements of the extracted ion beam by Trassl, Broetz et al. [6, 36] showed the temperature of the ion population to be relatively cold with a mean energy of 2 eV. Collisions between ions and electrons would lead to a thermalization of the plasma where all charged particles are part of the same Maxwellian temperature distribution. However, the observed ion charge states presented in the previous chapter indicate much higher electron energies.

8.1.1 Need for high energy electrons

One of the advantages of ECR ion sources is their ability to produce a wide variety of charge states as shown in the previous sections. The effective cross section for double or multiple ionization processes are at least two orders of magnitude smaller than effective cross sections for single ionization steps [33]. The production of ions with charge states higher than 1+ is therefore dominated by a step by step process [17] where a single electron is removed from an ion in each step rather than several electrons per step. Table 8.1 gives a first idea of the electron energies involved in the production of an argon ion beam from SWISSCASE containing all charge states up to Ar^{12+}:

This suggests the presence of electron energies above 600 eV. Ions with higher charge states are better confined in the ECR plasma than ions with lower charge

126 Chapter: Bremsstrahlung measurement

Charge state	Ionization potential [eV]	Beam ion abundance
Ar^+	15.759	0.346
Ar^{2+}	27.629	1.0
Ar^{3+}	40.74	0.523
Ar^{4+}	59.81	0.350
Ar^{5+}	75.02	0.211
Ar^{6+}	91.007	0.127
Ar^{7+}	124.319	$5.86 * 10^{-2}$
Ar^{8+}	143.456	$2.78 * 10^{-2}$
Ar^{9+}	422.44	$1.1 * 10^{-3}$
Ar^{10+}	478.68	$2 * 10^{-4}$
Ar^{11+}	538.95	$1.15 * 10^{-5}$
Ar^{12+}	618.24	$3.17 * 10^{-7}$
Ar^{13+}	686.09	-
Ar^{14+}	755.73	-
Ar^{15+}	854.75	-
Ar^{16+}	918.00	-

Table 8.1: Ionization potentials for different charge states of Argon [37]. Ion beam abundance: relative to maximally abundant charge state (Ar^{2+}). Argon charge states below the separator are not observed in the extracted ion beam of SWISSCASE but could be part of the SWISSCASE ECR plasma nevertheless due to the enhanced confinement for higher charge states.

states [11, 36] (see Chapter 3. Hence the ion charge state distribution of the extracted ion beam shows a lower charge state distribution than the ECR plasma itself. This suggests higher charge states than Ar^{12+} are present inside the ECR plasma calling for even higher electron energies than 600 eV.

Hohl et al. [17] have measured the electron temperature of the MEFISTO ECR plasma using an X-ray detector to capture the photons emitted from the ECR plasma. Despite the fact that the most abundant charge state of the MEFISTO ion beam is the single charge state, the outcome of the X-ray measurement was a hot electron population at a temperature of 2 keV. Additionally, Hohl et al. [17] stated that the fraction of these hot electrons is about 10% of all the ECR plasma electrons. However, in the SWISSCASE ion beam the most abundant charge states are 2+ for argon, 3+ for krypton and 5+ for xenon. The higher most abundant charge state of SWISSCASE suggests an electron temperature exceeding 2 keV, measured in the MEFISTO facility.

It is important to determine the hot electron temperature because the accurate knowledge of the electron temperature of the hot electron population in the SWISSCASE ECR plasma further allows the determination of the plasma density, the mean free path lengths, ionization cross sections for all observed charge states and the calculation of radiation power equilibrium as shown in Chapter 3.

In the next section the feasibility of a non-Maxwellian energy distribution and the potential for a high energy electron population is presented.

8.1.2 Runaway electrons and non Maxwellian energy distribution

To further understand the energy distribution of ECR electrons we consider the possibility of two different electron temperatures inside the same ECR plasma. For this we assume two different electron populations, one with a temperature of 2 eV and another with a non Maxwellian energy distribution and a mean energy of 10 keV. The cold electrons originate from the ionization process ot become part of the plasma. The hot electrons are plasma electrons that were energized in the ECR zone. The cold electron population is heated by the high energy electrons and the high energy electron population is cooled by the cold electrons in mutual collisions. This energy exchange happens by collisions between electrons of the cold and the high energy population. To estimate the energy transfer between the cold and the hot electron population we assume the high energy electrons to lose all of their kinetic energy as soon as they collide with another electron. In reality the hot electron can also collide with other hot electrons or with cold electrons without losing all their kinetic energy. Hence assuming total energy loss by one collision represents a conservative approximation with respect to the stability of the high energy population.

The energy loss of the high energy population can be described as follows:

$$\frac{P}{V} = f_{ee}\ E_{kin}\ n_h = n_c\ v\ \sigma\ E_{kin}\ n_h \tag{8.1}$$

$\frac{P}{V}$ is the energy loss rate of the high energy electrons per volume, f_{ee} is the collision frequency of the high energy electrons with the cold electrons, E_{kin} is the kinetic energy of the high energy electrons, n_h is the hot electron density, n_c is the cold electron density and v the electron velocity. σ is the effective collision cross section between the hot and the cold electrons. Geller [11] says that the dominating collision cross section is given by Spitzer collisions:

$$\sigma_{spitzer} = \left(\frac{e^2}{\epsilon_0\ E_{kin}}\right)^2 \frac{1}{2\pi}\ \ln\Lambda \tag{8.2}$$

$\sigma_{spitzer}$ is the collision cross section for Spitzer collisions, e is the electron charge, ϵ_0 the permittivity of vacuum, and $\ln\Lambda$ the Coulomb logarithm, a property weakly depending on the plasma parameters. $\ln\Lambda$ takes values between 5 and 30, in our case $\ln\Lambda$ is 24.65 [7]. Hence:

$$\frac{P}{V} = n_c\ v\ \left(\frac{e^2}{\epsilon_0\ E_{kin}}\right)^2 \frac{1}{2\pi}\ \ln\Lambda\ E_{kin}\ n_h = n_c\ n_h\ \frac{e^4}{\epsilon_0^2\ m_e\ v\ \pi}\ \ln\Lambda \tag{8.3}$$

m_e is the mass of an electron. By introducing $v = \sqrt{2E_{kin}/m_e}$ we assumed the kinetic electron energy to be small enough to apply non relativistic physics. From Eq. 8.3 we can see that the energy loss rate $\frac{P}{V}$ per volume is decreasing with increasing velocity and hence with increasing energy.

With increasing electron energy the Larmor radii increase. However the resonance condition is energy independent as long as the electron mass does not significantly change due to special relativity, i.e., as long as the electron energy

128   Chapter: Bremsstrahlung measurement

is significantly smaller than its rest mass energy equivalent of 512 keV. The energy input by the resonance mechanism is considered constant. This leads to a runaway effect, where electrons with a slightly higher kinetic energy than the cold bulk electron mean energy loose less energy while still benefiting from a constant power input. In this way a high energy electron population is created limited by effects like radiation loss, loss cone scattering and ionization collisions rather than energy transfer to the cold bulk electrons. This new dynamic equilibrium apparently stabilizes in a higher energy range which resembles a quasi-temperature [11, 3] of 2 keV for MEFISTO and 9-11 keVs for SWISS-CASE, as shown below, but can reach energies as high as 80 keV (Girard et al. [12] or 300 keV (Petty et al. [32]).

In the following section, attempts are presented to measure the hot electron temperature of SWISSCASE in a similar way as performed by Hohl et al. [17] at MEFISTO

8.2 Measurement setup

Hohl et al. [17] measured the Bremsstrahlung of MEFISTO in radial direction with respect to the confinement axis. Baru et al. [1] measured the Bremsstrahlung of MINIMAFIOS [27] in axial direction. Both measurement tactics, radial and axial, delivered an exponentially decaying intensity with increasing photon energy described further down this section. The Bremsstrahlung of the SWISSCASE ECR plasma can only be measured in axial direction due to the obstruction of the ECR plasma by the permanent magnets.

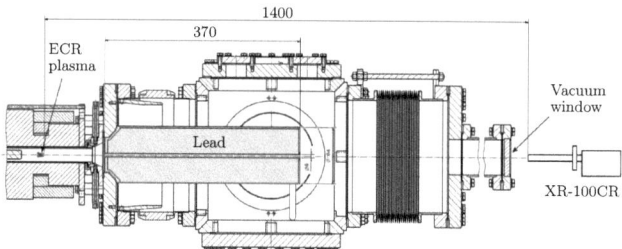

Figure 8.1: Setup for X-ray Bremsstrahlung measurement. A lead collimator (cyan) for X-ray fluorescence suppression is installed. The detector area measures 5 mm^2. All dimensions in mm.

Fig. 8.1 shows a section side cut of the measurement setup. A lead collimator is installed to suppress X-ray fluorescence induced by ECR electrons colliding with the UHV facility walls. The collimator features an aperture of 6 mm and restricts the view of the X-ray detector onto the ECR plasma. The ECR plasma is considered transparent for X-rays because the plasma cut-off frequency is limited by the much lower microwave input frequency. Consequently any X-rays emitted from the antenna tip (to the left side in Fig. 8.1) can be seen by the X-ray detector through the ECR plasma.

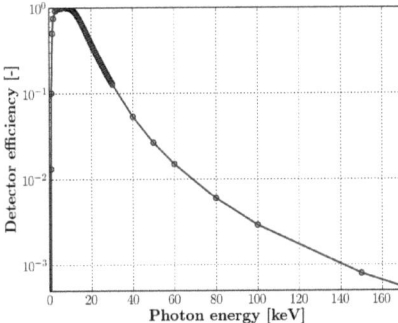

Figure 8.2: Efficiency of the XR-100CR detector. The detector efficiency gives the probability for an incident photon to be registered. Red dots represent the raw data provided by AMP TEK. The blue line represents a B-spline fit and is used for all subsequent calculations.

8.2.1 The X-ray detector

To detect the Bremsstrahlung from the ECR plasma the same Silicon X-ray detector from AMP TEK has been used which Hohl et al. [17] used to perform their X-ray measurements on MEFISTO. The XR-100CR detector features an energy resolution of 180 eV at 5.9 keV and came with a well-defined efficiency curve provided by AMP TEK. The Silicon detector head thickness measures 500 μm and is covered with a beryllium window of 12.7 μm thickness. The active detector area measures 5 mm^2. Fig. 8.2 gives the efficiency curve of the X-ray detector used for the acquisition of all subsequent X-ray spectra.

In the following section we discuss the necessary calibration of the X-ray detector.

8.2.2 Calibration of X-ray detector

The true photon flux Γ, the registered photon flux γ and the detector efficiency ϵ are related by Eq. 8.4.

$$\gamma = \Gamma \odot \epsilon \qquad (8.4)$$

$\odot$ stands for element-wise multiplication of both data arrays Γ and ϵ. Hence to obtain the true photon flux the measured photon flux array needs to be divided element wise by the efficiency data array. This calls for an exact co-alignment of both arrays in energy space. Radioactive ^{241}Am was used to obtain a proper co-alignment in energy space. ^{241}Am features a well known X-ray spectra from alpha decay into ^{237}Np. Table 8.2 summarizes the most relevant data about the radiation source used for calibration.

The radiation created by the alpha decay and the subsequent emission of X-ray from the decay product stimulate further X-ray emission from surrounding material of the radiation source. Silver is used for the backing and gold is the material in which the americium is electro deposited into. Hence line emission from those two elements are expected in addition to the afore mentioned emission energies of americium and neptunium. Fig. 8.3 shows an ^{241}Am spectrum obtained with the XR-100CR detector. Care has been taken to avoid pile up in

Isotope	$^{241}_{95}\text{Am}_{146}$	
Original activity	370 kBq	
Half life	432.2 years	
Decay product	$^{237}_{93}\text{Np}_{144}$	
Source diameter	11.1 mm	
Nature of active deposit	Electro deposited in gold matrix	
Backing	Silver	
Most prominent energies related to α-decay [keV]	Relative intensity [%]	Assignment
26.34	2.4	α
33.20	0.13	α
43.42	0.07	α
59.54	35.9	α
Most prominent energies related to X-ray emission from ^{237}Np [keV]	Relative intensity [%]	Assignment
13.95	9.6	Np $L_{\alpha 1}$
16.82	2.5	Np $L_{\beta 2}$
17.75	5.7	Np $L_{\beta 1}$
20.78	1.39	Np $L_{\gamma 1}$

Table 8.2: Details of the radioactive calibration source for the X-ray detector.

the counting rate of the silicon detector by choosing a distance of 2 m between the radiation source and the X-ray detector. No pile up effect was detected with this setup.

In Fig. 8.3 we identify the γ emission lines of ^{241}Am and line emission of neptunium, the decay product of ^{241}Am. Furthermore we see the $K_{\beta 1,3}$ emission line of silver at 22.1 and 24.9 keV. Silver is part of the americium source backing as indicated in Table 8.2. The observed silver emission line originates from X-ray fluorescence. In addition there are other peaks of line emission visible below 13 keV, which will be of more interest in the following sections where the actual results of the Bremsstahlung measurements are presented.

The obtained spectrum of ^{241}Am allows to precisely calibrate the X-ray detector in energy space which is necessary for the coupling with the efficiency curve provided by AMP TEK.

8.3 Post processing

In this section the effect of solid matter in the line of sight of the measured Bremsstrahlung is discussed. Disagreement between experimentally obtained absorption data and literature data disqualifies a limited energy region of the Bremsstrahlung measurement. However a significant energy range is identified where the attenuation, obtained by experiment, is validated by literature data. This restricted energy range is used to determine the ECR electron temperature

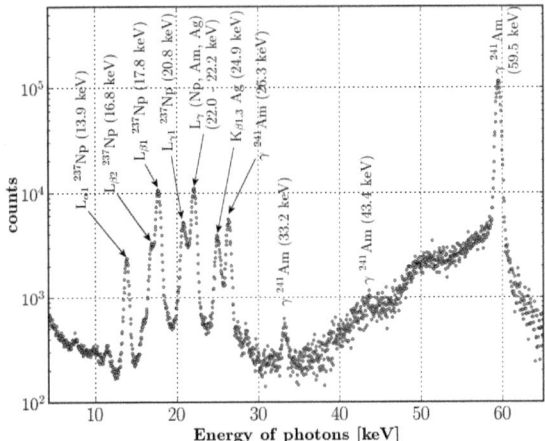

Figure 8.3: Spectrum of the described ^{241}Am radiation source, obtained with the X-ray detector XR-100CR. In addition to γ emission lines there are also X-ray emission lines from neptunium, americium and silver. Exposure time was 10 hours and the distance between source and detector measured 2 m. No pileup was detected.

in section 8.4.

8.3.1 Correction of attenuation effects

We are interested in a reliable magnitude representation of the registered X-ray radiation to extract the corresponding electron temperature as further detailed in Section 8.4. We need to consider solid material placed in the line of sight between the ECR plasma and the X-ray detector such as the vacuum window because solid material has great influence on the transmission of the measured X-rays and can be characterized by its energy dependent attenuation coefficient δ. In our measurements we used a vacuum window made of borosilicate. Figure 8.4 shows the attenuation coefficient of borosilicate according to literature [30].

The correction of the measured Bremsstrahlung spectrum by the literature data does not result in a useful Bremsstrahlung spectrum over the full range of photon energy as detailed in section 8.3.2. However the literature data is of importance for the identification of a restricted energy range where the literature data is in agreement with the experimentally obtained attenuation presented in the next section. The Bremsstrahlung spectrum of this restricted energy range is then used to determined the plasma electron temperature (see section 8.4).

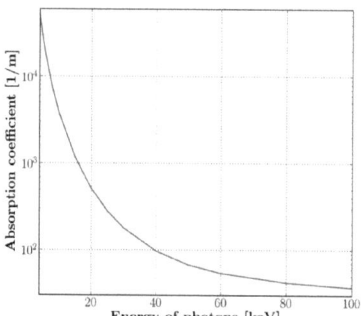

Figure 8.4: Attenuation coefficient of borosilicate according to literature [30]. Note the dynamic spread of 4 orders of magnitude which leads to strong absorption for photons of lower energies.

8.3.2 Validation of the Bremsstrahlung spectrum

Additional radiation, not originating from the ECR plasma, interference effects, scattered radiation, line emission outside the vacuum window or any absorption anomalies would falsify the Bremsstrahlung measurement to a certain extent.

A test is performed to compare the attenuation coefficient given by literature [30] and the attenuation coefficient obtained by an experiment described in this section. In the ideal situation with no unexpected falsifying radiation effect or interference phenomena, this test results in coincidence of literature and experiment. The experiment has shown, that this is not the case and there is indeed an unexplained effect, falsifying the Bremsstrahlung measurement for photon energies below 35 keV.

The following experiment is introduced to find differences between the attenuation of the vacuum window based on literature data and the photon flux measured with the X-ray detector. To obtain the attenuation coefficient of an unknown material an experimental setup can be used, provided a sufficiently continuous spectrum of photon energies is available in the energy range of interest. The principle is the comparison between two photon flux measurements, one flux measurement with a single vacuum window in place, a second flux measurement with a a second, additional but identical vacuum window in place.

Figure 8.5 presents the measurement setup to experimentally determine the attenuation coefficient δ of the two windows. The Bremsstrahlung measurement with original intensity I_1 is performed twice, once with one glass window in place, once with both glass windows in place. Comparing the two acquired spectra allows to determine the attenuation coefficient δ. Intensity I_2 is related to intensity I_1, δ and d by Eq. 8.5

$$I_2 = I_1 * e^{-\delta d} \tag{8.5}$$

And so is I_3:

$$I_3 = I_2 * e^{-\delta d} = I_1 * e^{-2\delta d} \tag{8.6}$$

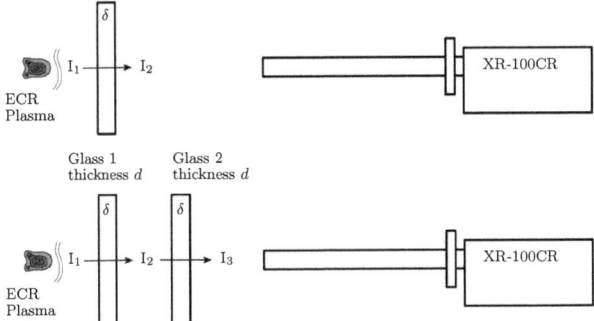

Figure 8.5: Schematics of setup to determine the attenuation coefficient δ of the used vacuum borosilicate glass window. The distance between the first glass and the detector is insignificant compared to the distance between the ECR plasma and the detector.

In our measurement I_1 is unknown, I_2 and I_3 are known. To access δ we divide Eq. 8.6 by Eq. 8.5.

$$\frac{I_3}{I_2} = e^{-\delta d} \quad (8.7)$$

Hence:

$$\delta = ln\left(\frac{I_2}{I_3}\right)/d \quad (8.8)$$

Note that, δ depends on photon energy and has to be computed separately for each energy channel of the obtained spectral data $I_2 \cdot \epsilon$ and $I_3 \cdot \epsilon$ with ϵ to be the energy dependent detector efficiency explained in the previous section. Any photon flux originating within or after the glass windows would lead to a wrong attenuation coefficient at the specific photon energy and deviate from literature data. This effect is indeed present and is explained in more detail here.

Figure 8.6 shows the attenuation effect obtained from the experiment compared to the attenuation effect from literature data presented in Figure 8.4. For energies between 35 keV and 70 keV the experimental data fits the literature data well. However for energies lower than 35 keV the experimentally obtained attenuation rather falls than it rises as the literature given attenuation does. For energies larger than 70 keV the experimentally obtained data suffers from low counts and correspondingly large standard deviation represented in a wide data point spread. The low count rate at higher energies lies in the nature of the ECR plasma emission which decreases with increasing photon energy [11, 14].

In the following section this deviation from literature data is investigated. An hypothesis for the deviation is formulated. In addition a valid energy range is defined, where the Bremsstrahlung spectrum is used for the determination of the electron temperature.

Figure 8.6: Attenuation effect obtained from the experiment in blue and attenuation effect calculated from literature data [30] in red. The experimentally obtained data fits the literature data well between energies of 35 keV and 70 keV. However for photon energies below and above this energy regime the experimental data significantly differs from the literature data. The data points above 70 keV suffer from low plasma emission photon number at higher energies.

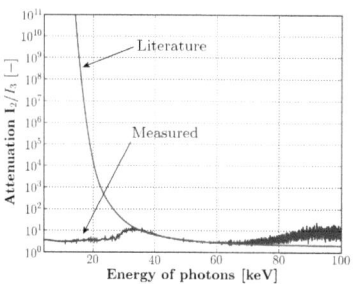

8.3.3 Discussion of attenuation effect

The attenuation coefficient obtained by the experiment does not agree with the attenuation coefficient given by literature [30] for certain energies. The experimental data is based on photon counts measured by the X-ray detector. Any additional radiation source beside the ECR plasma would inevitably falsify the measurement for both, the determination of the absorption coefficient and the Bremsstrahlung spectrum.

Several possibilities of measurement errors have been checked:

- An additional, falsifying photon flux has to be continuous in order to simulate the continuous difference in the experimental and the literature given attenuation. Consequently line emission or X-ray fluorescence from any source, at the detector tip or the environment, cannot alone be responsible for the observed effect.

- Measurements with the ECR plasma being switched off have shown that the noise of the X-ray detector lies below 10 counts per hour and can be safely considered as insignificant. Hence the effect has to be stimulated by the ECR plasma.

- The ECR plasma is very bright in ultra violet (UV) light [11]. UV radiation could interfere with semiconductors and produce an erroneous detector output. However the X-ray detector tip is shielded by a beryllium window of 12.7 μm thickness. This window is considered to be opaque with respect to the wave length of UV light [17].

- Measurements were performed with active microwave power but with a feed gas pressure ($5*10^{-8}$ mbar) too low to allow plasma operation. This allowed to check any interference between the microwave activity and the measurement setup without the presence of an ECR plasma. During these measurements the X-ray detector showed the well known back-ground count rate of 10 counts per hour (see above).

- The cooling pump motor, the laboratory ventilation and nearby devices were switched on and off to investigate any interference between the measurement setup and sharp current interruptions of circuits or motor brushes. Again the X-ray detector showed the back-ground count rate of 10 counts per hour.

However the difference in the measured attenuation and the literature based attenuation may be explained by additional photons created at the detector front face. These photons had to be in the actual energy range of up to 35 keV rather than UV energies. Such an additional photon flux would explain the deviation of the experimentally obtained attenuation coefficient and the attenuation given by literature at respective photon energies. The additional photons originating from the detector tip would simulate an apparently higher vacuum window transmission and a lower attenuation because they are not attenuated by the vacuum window.

There is another indicator suggesting that additional radiation is emitted from the X-ray detector. Emission lines of silver can be observed in both the Bremsstrahlungs spectrum and the calibration procedure using ^{241}Am. To visualize this effect figure 8.7 shows an overlay of a ^{241}Am spectrum and a plasma operation spectrum, uncorrected for attenuation.

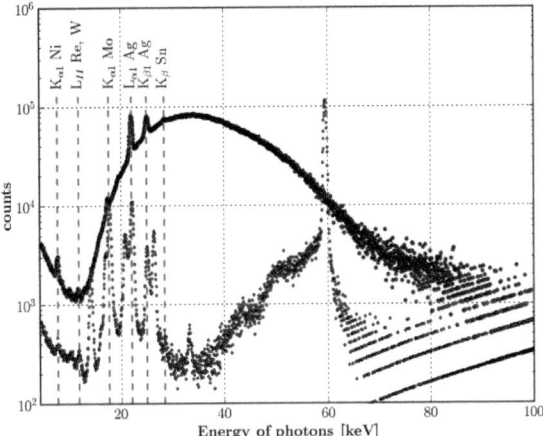

Figure 8.7: Overlay of argon plasma Bremsstrahlungs spectrum in blue and ^{241}Am spectrum in red. The ^{241}Am has been produced with no solid matter in between radiation source and X-ray detector. Vertical green dashed lines indicate emission lines occurring in both spectra indicating the presence of the respective material in the detector rather than the radiation source or the plasma.

In figure 8.7 emission lines of rhenium, tungsten, nickel, molybdenum, tin and silver can be seen. Rhenium, tungsten, nickel and molybdenum are common components of steel and therefore cannot be assigned to a specific location without doubt. However silver and tin are usually not present in the construction material of the vacuum facility and tin is used as soldering metal on the IC board of the detector chip. Silver is part of an internal collimator of the X-ray detector.

Both the presence of tin and silver emission lines indicate the additional X-ray activity of the detector and thereby supports the hypothesis that additional radiation is originating from the detector tip, falsifying the Bremsstrahlung measurement below 35 keV.

At this point the exact nature of the observed effect is unknown. However for energies between 35 keV and 70 keV the experiment is in excellent agreement with literature data. Hence the useful energy regime is identified and can be used for the determination of the electron temperature as shown in the next section.

8.4 Determination of electron temperature

Assuming a Maxwellian kinetic energy distribution for the hot electrons, the intensity $I(E_{photon})$ of the emitted Bremsstrahlung as a function of the photon energy E_{photon} can be written as: ([3, 12, 19])

$$I(E_{photon}) = I_0 \, \exp\left(-\frac{hf}{k_B T_e}\right) = I_0 \, \exp\left(-\frac{E_{photon}}{k_B T_e}\right) \tag{8.9}$$

I_0 is the radiation intensity at the energy $E_{0photon}$ and basically only scales the whole function along its magnitude. h is the Planck constant, f the photons angular frequency, k_B the Boltzmann constant and T_e the hot electron temperature. Plotted in a logarithmic scale the intensity $I(E_{photon})$ appears as a straight line with a slope $a = -1/k_B T_e$. This allows to estimate the temperature of the emitting electron population as a first approximation.

The procedure to determine the electron temperature from the X-ray spectrum assumes a Maxwellian energy distribution of the ECR plasma electrons. Due to the presence of a strong magnetic field and due to the energy decreasing collision cross section between different energy populations the development of an electron ensemble with a Maxwellian energy distribution is unlikely. Hence this approximation of the plasma electron temperature is only a first order approximation used to compare mere orders of magnitude between different ECR Bremsstrahlung measurements [11, 17, 14].

Figure 8.8 shows a Bremsstrahlung spectra of MEFISTO performed by M. Hohl in 2002 [16] to give an idea of the expected exponential decay in X-ray intensity with increasing photon energy.

figure 8.9 shows two Bremsstrahlung spectra obtained from an argon plasma of SWISSCASE. Both spectra were corrected by the attenuation factor based on literature data [30] because literature data and experimental data agree between 35 keV and 70 keV, the energy range which is used for the determination of the electron temperature. In addition the literature data provides continuous and precise data also for larger energies, where the experiment suffers from low count rates resulting in low resolution of the experimentally obtained absorption coefficient. A dashed green line marks the 35 keV limit below which literature

Chapter: Bremsstrahlung measurement 137

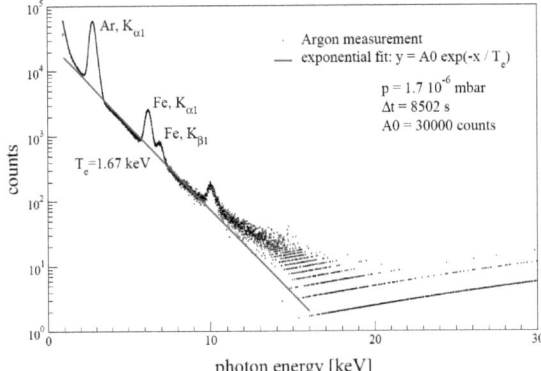

Figure 8.8: Bremsstrahlungs spectrum generated by an argon plasma in MEFISTO. The slope x of the exponential function represents a temperature of $T_e = 1.7$ keV. Picture: Ph.D. thesis from M. Hohl 2002 [16].

data and the attenuation effect obtained experimentally do not agree anymore due to the unknown measuring effect discussed in section 8.3.3.

In addition, both spectra were obtained with the same plasma parameter settings except that the feed gas pressure to be $1.8*10^{-4}$ mbar for the measurement shown in blue and $5*10^{-5}$ mbar for the measurement shown in green.

In figure 8.9 we can see both exponential decay functions fit the measured data exactly for energies higher than 35 keV. The exponential fit deviates from the measured spectrum at the same energy. This again supports the thesis, the collected data below 35 keV to be subject of an unknown measurement effect and not part of the actual plasma X-ray spectrum.

8.5 Conclusion

In the validated energy range between 35 keV and 70 keV the Bremsstrahlung spectrum shows a constant intensity slope, expected by theory (8.9). The temperature derived from this slope is 9 keV for an argon plasma operated at $1.8*10^{-4}$ mbar, 10.88 GHz of input frequency and 95 W of input power. For the same frequency and power setting but a lower feed gas pressure of $5*10^{-5}$ mbar the corresponding derived electron temperature is 11 keV.

This result is in excellent agreement with theory which stipulates higher electron energies for lower feed gas pressures due to the lower particle densities and the correspondingly longer mean free path lengths [11]. Given longer mean free path lengths the electrons undergo a longer acceleration process yielding higher electron energies and temperatures.

The average electron temperature resulting from the experiment is 10 keV and is further used for the calculation of plasma parameters in chapter 'Electron Cyclotron Resonance'.

138　　　　　　　　　　Chapter: Bremsstrahlung measurement

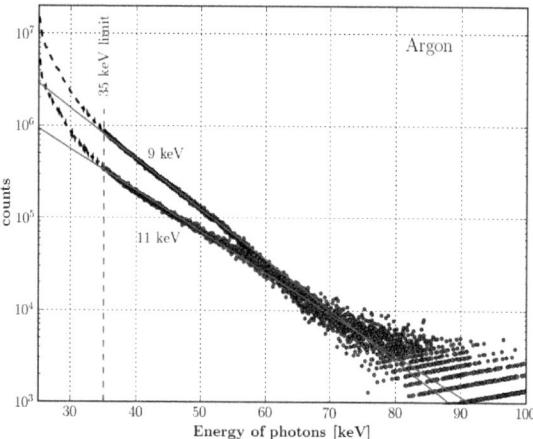

Figure 8.9: Bremsstrahlung spectrum of two different ECR plasmas. One is obtained at a feed gas pressure of $1.8*10^{-4}$ mbar, shown in blue. The other is obtained at a feed gas pressure of $5*10^{-5}$ mbar, shown in green. $f = 10.88$ GHz, $P = 95$ W for both spectra. Two exponential decay functions (red lines) represent electron temperatures of 9 keV for the data in blue and 11 keV for the data in green respectively. The 35 keV limit to where the measurement is correct is shown by a vertical dashed line in green. Note, the exponential fit deviates from the measured spectrum approximately at the same point where the 35 keV limit intersects the spectrum.

Chapter 9

Summary

An new ECR ion source, SWISSCASE (**S**olar **W**ind **I**on **S**ource **S**imulator for the **CA**libration of **S**pace **E**xperiments), has been realized at the University of Bern. The facility will be used to calibrate space borne instrumentation.

The new ion source SWISSCASE needs to be integrated into CASYMS and will be operated parallel to the existing filament based electron collision ion source. SWISSCASE has to produce ion currents of highly charged ions six to eight orders of magnitude more intense than MEFISTO or the CASYMS filament ion source. Because it will eventually be operated on a high voltage terminal the new ion source is favored to require low feed power.

As the other ion sources of the University of Bern, the new ion source will be used to test and calibrate space borne experiments under ultra high vacuum conditions. For this, the ion source must keep the vacuum contamination in the target chamber from the plasma operation below 10^{-8} mbars.

9.1 ECR Theory

It was shown that the performance of the magnetic confinement can be expressed by single particle motion widely found in literature (Chen, Geller). Initial phase differences between different ECR electrons do not impede the overall high energy electron population because all electrons get accelerated up to large energies (10 keV). Diffusion is of moderate interest because it relies on multiple scattering which disqualifies the hot electrons and expels the cold electrons and ions from the plasma.

However mean free paths are long enough for the hot electrons to build up a negative plasma potential to improve the ion confinement. This is necessary to enable a plasma where low charged ions can collide several times with hot electrons to undergo the described step by step ionization process and build up the highly charged ion population.

The increase in operation frequency from MEFISTO to SWISSCASE significantly improves the critical electron density by a factor of $\sim$20. In addition the SWISSCASE plasma is more compact and features, defined by the ECR zone, a volume 8.8 times smaller than MEFISTO. This leads to a similar overall electron content but a significantly larger ratio of electron content to ECR zone surface area in favor for SWISSCASE.

9.2 Realization of source elements

SWISSCASE is a full permanent confinement magnet ECR ion source operated at 10.88 GHz for an argon plasma. The facility is ultra high vacuum capable. A slidable antenna and a short allow impedance matching. A movable puller electrode and Einzel lense assembly enable extraction geometry adjustments for optimal ion beam formation. A baffle has been implemented which shields the high voltage insulation ceramics from surface coating of redeposited sputtering material.

Both electrical and mechanical operation settings can be manipulated remotely and without either shutting the extraction voltage nor breaking the vacuum. However if a vacuum venting is desired to exchange inner components, the facility is able to reach UHV conditions within 24 hours.

The ion beam output is analyzed by a 90° mass separation magnet and measured by a sophisticated two stage Faraday cup which suppresses secondary electron emission.

9.3 FEM MEFISTO and FEM SWISSCASE

The numerical simulation of MEFISTO and SWISSCASE revealed the detailed magnetic structure of the ECR plasmoid. The magnetic confinement of future ECR ion sources can be designed with high accuracy with respect to location and field structure.

In both, the simulation of SWISSCASE and MEFISTO, magnetic field and electron density maps were correlated with common sputtering and surface coating patterns found in hexapole ECR ion sources, which verifies the simulation concept and supports its quality. In MEFISTO the magnetic field closely resembles the observed ion coating patterns. However, in SWISSCASE, the magnetic field simulation has to be complemented with electron density maps to reveal the same correlation.

In SWISSCASE, the hot ECR electrons and the plasma ions are closely related in both, position and velocity space, due to the strong magnetic field and the resulting similar larmor radii of hot ECR electrons and cold plasma ions. This can significantly simplify future computations of ion distribution inside ECR systems, such as scientific ion sources, ion implanters for semi conductor manufacturing and ion engines for space propulsion.

9.4 Performance, characterization and fulfilled requirements

SWISSCASE delivers a broad spectrum of different charge states up to Ar^{12+}, Kr^{13+} and Xe^{12+} respectively. The ion charge state distribution is continuous with a maximum ion current at charges higher than one for all tested gases except the very light ones, carbon and oxygen. This supports the high ECR electron temperature of 10 keV presented in Chapter 8.

The broad presence of charge states throughout a large parameter field greatly simplifies the operation and the tuning of SWISSCASE. The extraction of ions from the ECR plasma leads to sputtering on the puller electrode.

The sputtering results in wear and degrades the puller electrode. SWISSCASE has been operated approximately 200 hours and 400 hours with different operation gases. The wear patterns on the electrodes are well visible but did not lead to a decrease in ion beam performance over the test period. Hence, despite the wear on the puller electrode SWISSCASE can be operated well beyond the tested time scale. A significant decrease in ion beam performance would give due notice of a puller electrode eventually in progressed sputtering after many hundreds of hours of operation time.

SWISSCASE fulfills all hard and soft requirements defined in Chapter 2 to the specified values as shown in Table 9.1.

Requirement	Value required	hard /soft	Value achieved
Ion charge states	Ar^{1+} to Ar^{10+}	hard	Ar^{1+} to Ar^{10+}
Ion charge states	Ar^{11} to Ar^{n+}	soft	Ar^{11+} to Ar^{12+}
Ion current @ Ar^{8+}	$1.78*10^{-1}$ μA	hard	2 μA
Input power	< 750 W, 1 phase	soft	600 W, 1 phase
UHV capability at target	10^{-8} mbar	hard	10^{-8} mbar
Material cost	250'000 CH Fr.	hard	200'000 CH Fr.

Table 9.1: All requirements are fulfilled.

9.5 Bremsstrahlung measurement

Between a photon energy of 35 keV and 70 keV the Bremsstrahlung measurement revealed a hot electron temperature of 9 keV for an argon plasma operated at $1.8*10^{-4}$ mbar, 10.88 GHz of input frequency and 95 W of input power. A different hot electron temperature of 11 keV was obtained with an argon plasma operated at $5*10^{-5}$ mbar at the same frequency and the same microwave input power. The difference in hot electron temperature is in excellent agreement with literature which predicts hotter electrons for a lower feed gas pressure [11]. The average hot electron temperature resulting from these two measurements is 10 keV which is in good agreement with literature.

9.6 Outlook

SWISSCASE is ready for integration into CASYMS with the acquired knowledge about the tested gases. It can also be tested further with other gases such as organics like propane and butane. This would reveal its ability to ionize molecules without disintegration.

A microwave model of the ECR plasma would allow to determine the microwave impedance. Finite element modeling with commercial software such as Opera by Vectorfields allows to model the remaining facility. Combining both models, the plasma model and the facility model, enables the computation of the absorbed and the reflected microwave power in the plasma.

Sweeping the microwave frequency allows sweeping of the ECR zone location (see Chapter 3) which results in new ion current output data. The output frequency of the implemented microwave generator can be controlled via terminal by an external signal generator. The resulting periodic compression and expansion of the ECR zone can lead to several interrelated effects such as:

- improved neutrals supply for the ECR zone
- weakened electron and ion confinement
- shorter ion confinement times
- improved extraction of highly charged ions due to weakened ion confinement
- new plasma effects

Due to the hardware already in place, experimental runs would verify whether this results in a beneficial overall effect for highly charged ion production.

The puller electrode can be exchanged with a version of different length, thereby changing the distance between extraction and first Einzel-lens. Simulations with Opera3D and experimental runs could reveal possible improvements in the formation of highly charged ion beams.

9.7 Conclusion

To conclude, SWISSCASE is a new, very successful, high performance ECR ion source with a full permanent magnet confinement. SWISSCASE reliably delivers ion beams of highly charged ions on demand with highest reproducibility and excellent vacuum quality.

In addition, the findings of this thesis allow the accurate and reliable design of magnetic confinements and the computation of electron and ion distribution inside ECR ion sources. This furthers the efficient fabrication of a wide range of future electron cyclotron resonance applications to meet the demand of modern charged particle related science fields, semiconductor industries and spacecraft propulsion.

Bibliography

[1] C. Baru, P. Briand, A. Girard, G. Melin and G. Briffod, Hot electron studies in the Minimafios ECR ion source, Review of Scientific Instruments, Volume 63, Number 4, April, p.2844-2846, 1992.

[2] W. Beitz, K.H. Grote, Dubbel Taschenbuch fuer den Maschinenbau, 20. Auflage, Springer, 2001.

[3] K. Bernhardi and K. Wiesemann, X-ray Bremsstrahlung measurements on an ECR discharge in a magnetic mirror, Plasma Physics 24, p.867-884, 1982.

[4] M. Bodendorfer, Field structure and electron life times in the MEFISTO Electron. Cyclotron Resonance Ion Source, Nuclear Instrumentes and Physics research B, 266, p.820-828, 2008.

[5] W. H. Bostick, Experimental Study of Ionized Matter Projected across a Magnetic Field, Physics Review 104, p.292-299, 1956.

[6] F. Broetz, Ph.D. thesis, Weiterentwicklungen und Untersuchungen an einer herkoemmlichen 14 GHz und verschiedenen vollpermanenten 9-10.5 GHz ElektronZyklotronResonanz (EZR)Ionenquellen, University of Giessen, Germany, 2000.

[7] F. F. Chen, Introduction to Plasma Physics and Controlled Fusion, Volume 1: Plasma Physics, Second Edition, Plenum Press (1984).

[8] E.L. Cussler, Diffusion, Mass Transfer in Fluid Systems, Second Edition, Cambridge University Press, 2000.

[9] P. C. Efthimion, RF PLASMA SOURCE FOR HEAVY ION BEAM CHARGE NEUTRALIZATION,Philip C. Efthimion, Plasma Physics Laboratory, Princeton University, Princeton, NJ 08543, Proceedings of the 2003 particle accelerator conference.

[10] R. Friedlein; Experimental study of the hot and warm electron populations in an electron cyclotron resonance argonoxygenhydrogen plasma Cyclotron Resonance Ion Source, Nuclear Instrumentes and Physics research B, Volume 266, Issue 5, pages 820-828, March, 2008.

[11] R. Geller, Electron Cyclotron Ion Sources and ECR Plasmas, Institute of Physics Publishing Bristol and Philadelphia (1996).

[12] A. Girard, P. Briand, G. Gaudart, J.P. Klein, F. Bourg, J. Debernardi, J.m. Mathonnet, G. Melin and Y. Su, The Quadrumafios electron cyclotron resonance ion source: presentation and analysis of the results Rev. Sci. Instrum. 65, p.171417, 1994.

[13] A.G. Ghielmetti, H. Balsiger, R. Baenninger, P. Eberhardt, J. Geiss and D.T. Young, Calibration system for satellite and rocket-borne ion mass spectrometers in the energy range from 5 eV/q to 100 keV/q, Review of Scientific Instruments, 54, p.42536, 1983.

[14] K.S. Golovanivsky and G. Melin, A study of the parallel energy distribution of lost electrons from the central plasma of an electron cyclotron resonance ion source, Review of Scientific Instruments, Volume 63, Number 4, p.2886, April, 1992.

[15] K. Halbach, Design of permanent magnet multipole magnets with oriented rare earth cobalt materials, Nuclear Instruments and Methods 169, p. 1-10, 1980.

[16] M. Hohl, Ph.D. thesis, Design, setup, characterization and operation of an improved calibration facility for solar plasma instrumentation, University of Bern, Switzerland, 2002.

[17] M. Hohl, Investigation of the density and temperature of electrons in a compact 2.45 GHz electron cyclotron resonance ion source plasma by x-ray measurements, PLASMA SOURCES SCIENCE AND TECHNOLOGY, Number 14, p.692699, 2005.

[18] J.D.Huba, NRL PLASMA FORMULARY 2007, Beam Physics Branch, Plasma Physics Devision, Naval Research Laboratory, Washington, DC 20375.

[19] I. H. Hutchinson, Principles of Plasma Diagnostics, Cambridge University Press, 1987.

[20] J. D. Jackson, Classical Electrodynamics, Third Edition, John Wiley and Sons Inc., 1998.

[21] R. Karrer, Ph.D. thesis, Ion optical calibration of the PLASTIC sensor and STEREO, University of Bern, Switzerland, April, 2007.

[22] M. Liehr, Ph.D. thesis, University of Giessen, Germany, 1992.

[23] M. Lier, R. Trassl, M. Schlapp, E. Salzborn, A low power 2.45 GHz ECR ion source for multiply charged ions, Review of Scientific Instruments, 63 April 2541, p.2541-2543, 1992.

[24] W. Lotz, Electron-impact ionization cross-sections and ionization rate coefficients for atoms and ions from hydrogen to calcium, Zeitschrift fuer Physik, 216, p.241-247, 1968.

[25] R.E. Marrs, S.R. Elliot, D.A. Knapp, Lawrence Livermore National Laboratory, Production and Trapping Hydrogenlike and Bare Uranium Ions in an Electron Beam Ion Trap, Physical Review Letters, Vol. 72, Number 26, p.4082-4085, June 27th, 1994.

[26] A. Marti et al., Calibration facility for solar wind plasma instrumentation, Review of Scientific Instruments, Volume 72, Number 2, p.1354-1360, February, 2001.

[27] G. Melin et al., Proc. 10th Intern. Workshop ECR Ion Sources, Oak Ridge, report ORNL CONF-9011136, 1, 1990.

[28] Microwave Power Inc., 3350 Scott Blvd. Building 25, Santa Clara, CA 95054, Phone: (408) 727-6666, Fax: (408) 727-2246, www.microwavepower.com, info@microwavepower.com

[29] A. Mueller, E. Salzborn, R. Frodl, R. Becker, H. Klein and H.P. Winter, Absolute ionization cross sections for electron incidents on O^+, Ne^+, Xe^+, and Ar^{i+} (i = 1, ..., 5) ions , Journal of Physics, B 13, p.1877, 1980.

[30] National Institute of Standards and Technology, $http://physics.nist.gov/PhysRefData/Ionization/$.

[31] V.P. Ovsyannikov, G. Zschornack, Precision adjustment of an ECR plasma chamber relative to the magnetic confinement field using strong focused electron beams, Nuclear Instruments and Methods in Physics Research, A 416, p.18-22, 1998.

[32] C.C. Petty, D.L. Goodman, D.L. Smatlak and D.K. Smith, Confinement of multiply charged ions in a cyclotron resonance heated mirror plasma, Phys. Fluids, B 3, p.70514, 1991.

[33] S. Rachafi, Absolute cross section measurements for electron impact ionization of Ar^{7+}, J. Physics, B: At. Mol. Opt. Phys. 24, p.1037-1047, UK, 1991.

[34] G. D. Shirkov, A classical model of ion confinement and losses in ECR ion sources, PLASMA SOURCES SCIENCE AND TECHNOLOGY, page 250, Number 2, 1993.

[35] G. D. Shirkov, Electron and ion confinement conditions in the open magnetic trap of ECR ion sources, CERN-PS-94-13, 11th June, 1994.

[36] R. Trassl, Ph.D. thesis, Entwicklung "vollpermanenter" EZR Ionenquellen und Untersuchung des Ladungsaustausches in Stoessen zwischen 4-fach geladenen Wismut-Ionen, University of Giessen, Germany, 1999.

[37] R. C Weast, Handbook of Chemistry and Physics, 70th Edition, CRC Press. Inc.

[38] D.C. Wilcox, Basic Fluid Mechanics, first edition, second printing 1998, ISBN 0-9636051-3.

[39] P. Wurz, A. Marti, P. Bochsler, New test facility for solar wind instrumentation, Helvetica Physica Acta 71, p.23-24, 1998.

[40] Wutz, Adam, Walcher, Theorie und Praxis der Vakuumtechnik, 3. Auflage, 1986. ISBN 3-528-14884-5.

List of Figures

1.1 PLASTIC instrument and Delta II launch 3
1.2 Size comparison between SWISSCASE and JET 4
1.3 3D Overview of SWISSCASE 5
1.4 Top view of SWISSCASE 6

3.1 Cyclotron motion of an electron 14
3.2 Magnetic mirror 17
3.3 The magnetic flux is conserved 18
3.4 Loss cone 18
3.5 Electron cyclotron resonance 20
3.6 Electron trajectories in resonance with a magnetic field 21
3.7 Kinetic electron energy with respect to time 22
3.8 Kinetic electron energy with respect to the distance 23
3.9 Surplus electrons 28
3.10 Cylinder cut of SWISSCASE 29

4.1 3D overview of SWISSCASE realized within this thesis 35
4.2 Top view section cut through SWISSCASE 36
4.3 Side view section cut through SWISSCASE 36
4.4 Section cuts of the magnet assembly used in SWISSCASE 37
4.5 Magnetic field along z-axis of SWISSCASE 39
4.6 Demagnetization curves of VACODYM 677 HR 40
4.7 3D overview of the SWISSCASE magnetic field 41
4.8 Overview of the microwave system 43
4.9 Composition of power measurement system 45
4.10 Measured power output of the microwave generator 46
4.11 Schematic of a ferrit core circulator 46
4.12 Section cut of impedance matching 47
4.13 Section cut of extraction setup 48
4.14 Realization of extraction assembly 49
4.15 Photographs of puller electrodes 49
4.16 Photographs of puller electrode tips 50
4.17 A section cut of baffle 51
4.18 Composition of Einzel lenses 52
4.19 Helium spectrum acquired with original extraction setup 54
4.20 Vacuum and gas flow setup 56
4.21 Argon spectrum of SWISSCASE 57
4.22 High voltage parts 60

List of Figures

4.23	Potential distribution in the extraction region	61
4.24	Electric field at the puller electrode	62
4.25	3D view of high voltage parts	63
4.26	Potential distribution along z-axis	63
4.27	Sketch for calculation of magnetic path integral	65
5.1	Full permanent magnet arrangement	69
5.2	Finite element model of MEFISTO	70
5.3	Magnetic field along the beam axis	71
5.4	Cut planes in the 3D model	71
5.5	MEFISTO: magnetic field in plane A	72
5.6	MEFISTO: detailed view of plane A	72
5.7	MEFISTO: magnetic field in plane b	73
5.8	MEFISTO: detailed view of plane B	73
5.9	MEFISTO: magnetic field in plane C	74
5.10	MEFISTO: detailed view of plane C	74
5.11	Plasmoid MEFISTO	75
5.12	Magnetic field density in a plane parallel to plane C	76
5.13	Initial velocities parallel to field lines	77
5.14	Initial velocities perpendicular to field lines	78
5.15	Histogram of particle life	78
5.16	Velocity versus trajectory length, simulation A	79
5.17	Velocity versus trajectory length, simulation B	80
5.18	Current density maps of MEFISTO	82
6.1	Cut planes	89
6.2	Overlay of magnetic fields	89
6.3	SWISSCASE: magnetic field in plane A	90
6.4	SWISSCASE: detailed view of plane A	90
6.5	SWISSCASE: magnetic field in plane B	91
6.6	SWISSCASE: detailed view of plane B	91
6.7	SWISSCASE: magnetic field in plane C	92
6.8	SWISSCASE: detailed view of plane C	92
6.9	Plasmoid SWISSCASE	93
6.10	Electron current density maps of SWISSCASE	94
6.11	Photograph of the microwave antenna	96
7.1	Extracted ion current of Ar^{8+}	100
7.2	Microwave antenna position	101
7.3	Schematics of microwave situation	102
7.4	Optimum microwave antenna positions	103
7.5	Setup for ion current measurement	105
7.6	Section cut of two stage Faraday cup	107
7.7	3D view of two stage Faraday cup	107
7.8	Potential distribution	108
7.9	Quarter of finite element model	108
7.10	Spectrum of argon plasma	110
7.11	Comparison small aperture, large aperture	111
7.12	Spectrum of argon plasma, optimized for Ar^{12+}	112
7.13	Krypton plasma spectrum, optimized for Kr^{8+}	113

List of Figures

7.14 Magnified view of the peaks Kr^{2+} and Kr^{3+} 114
7.15 Krypton spectrum optimized Kr^{13+}. 115
7.16 Xenon plasma spectrum, optimized for Xe^{6+} 116
7.17 Xenon spectrum optimized for Xe^{12+}. 117
7.18 Carbon dioxide spectrum optimized for C^{3+} 118
7.19 Carbon dioxide spectrum optimized for C^{5+} 119
7.20 Extracted ionization power for each charge state 121

8.1 Setup for X-ray Bremsstrahlung measurement 128
8.2 Efficiency of the XR-100CR detector 129
8.3 X-ray spectrum of ^{241}Am 131
8.4 Attenuation coefficient of borosilicate, literature 132
8.5 Setup to determine the attenuation coefficient 133
8.6 Attenuation effect obtained from the experiment 134
8.7 Overlay of argon plasma Bremsstrahlungs spectrum and ^{241}Am spectrum 135
8.8 Argon Bremsstrahlungs spectrum, uncorrected 137
8.9 Bremsstrahlung spectrum of two different argon ECR plasmas . . 138

List of Tables

1.1	Elements in the solar photosphere, the slow and the fast solar wind	1
2.1	Instrument currents	8
2.2	Requirements	9
3.1	Summary of cross sections	27
3.2	Summary of design and optimal plasma operation parameters	32
4.1	Comparison of High-B-field confinements	39
4.2	Requirements for the microwave generator	44
4.3	Summary of purchase options	44
4.4	Specification of microwave generator	45
4.5	Molecular flow	55
4.6	Summary of vacuum values	59
4.7	Potential ranges of interest	60
4.8	Tested voltage limits	64
4.9	Entities, values and units used path integral	65
5.1	Characteristics of the finite element model	70
5.2	Breakdown of the trajectory length	80
5.3	summary of cluster separation	80
6.1	Spitzer collision cross sections	86
6.2	Mean free path lengths	86
6.3	Characteristics of the finite element model	88
6.4	Composition of surface coating	95
7.1	Parameter settings and logging of SWISSCASE	106
7.2	Stable argon isotopes	109
7.3	Summary peak currents	111
7.4	Stable stable krypton isotopes	113
7.5	Stable stable xenon isotopes	116
7.6	Stable oxygen and carbon isotopes	118
7.7	Step wise ionization energy	121
8.1	Ionization potentials for different charge states	126
8.2	Details of the radioactive calibration	130
9.1	All requirements are fulfilled	141

Chapter 10
Acknowledgments

This thesis would not have been possible without the extensive support, the patient help and the good spirit of my supervisors Prof. Dr. Herbert Shea and Prof. Dr. Kathrin Altwegg. Thank you very much. In addition I thoroughly thank

- Prof. Dr. Peter Wurz for the infinite number of suggestions, tips and explanations in all parts of this thesis and both publications.
- Prof. Dr. Peter Bochsler for enabling this work in the first place.
- Prof. Dr. Erhard Salzborn for his extensive support and the recommendations for a 10 GHz ECR ion source.
- Dr. Lisa M. Blush for her help in plasma physics and many very interesting discussions.
- Dr. Martin Wieser for numerous tips in constructing a low noise level Faraday cup.
- Dr. Reto Karrer for important tips and interesting discussions about the PLASTIC flight hardware
- Adrian Etter for his extensive support in the laboratory and his good nature.
- Dr. Daniele Piazza, Harry Mischler, Stefan Graf, Joseph Fischer, Martin Sigrist, Michael Gerber and Beat Zahnd from the engineering department for the construction of the new UHV facility and the Faraday cup.
- Juerg Jost, Roland Nussbaum and Martin Neuenschwander from the electronics department for the realization of many subsystems for SWISS-CASE.
- Dr. Urs Jenzer for his patient IT support and many interesting discussions.
- Prof. Dr. Urs Moser for his help in high energy physics and the Bremsstrahlung measurement.
- my parents for their long lasting support and their rock solid belief in me.

- all my friends for many good discussions and beautiful memories during the last four years.

This work was supported by the Swiss National Science Foundation.

I want morebooks!

Buy your books fast and straightforward online - at one of world's fastest growing online book stores! Environmentally sound due to Print-on-Demand technologies.

Buy your books online at
www.morebooks.shop

Kaufen Sie Ihre Bücher schnell und unkompliziert online – auf einer der am schnellsten wachsenden Buchhandelsplattformen weltweit! Dank Print-On-Demand umwelt- und ressourcenschonend produziert.

Bücher schneller online kaufen
www.morebooks.shop

KS OmniScriptum Publishing
Brivibas gatve 197
LV-1039 Riga, Latvia
Telefax: +371 686 204 55

info@omniscriptum.com
www.omniscriptum.com

Printed by Books on Demand GmbH, Norderstedt / Germany